Surveillance of Drinking Water Quality in Rural Areas

Dr Barry Lloyd
Robens Institute of Health and Safety, University of Surrey, Guildford
Dr Richard Helmer
Division of Environmental Health, World Health Organization, Geneva

Surveillance of Drinking Water Quality in Rural Areas

Published on behalf of the World Health Organization and the United Nations Environment Programme

Copublished in the United States with
John Wiley & Sons, Inc., New York

Longman Scientific & Technical
Longman Group UK Limited
Longman House, Burnt Mill, Harlow
Essex CM20 2JE, England
and Associated Companies throughout the world

Copublished in the United States with
John Wiley & Sons Inc., 605 Third Avenue, New York, NY 10158

First published 1991

British Library Cataloguing in Publication Data
Lloyd, Barry
Surveillance of drinking water quality in rural areas.
1. Rural regions. Drinking water. Surveillance
I. Title II. Helmer, Richard *1943–* III. World Health Organization IV. United Nations Environment Programme
363.61

ISBN 0–582–06330–2

Library of Congress Cataloging in Publication Data
Lloyd, Barry.
Surveillance of drinking water quality in rural areas / Barry Lloyd, Richard Helmer.
p. cm.
'Published on behalf of the World Health Organization and the United Nations Environment Programme.'
Includes bibliographical references and index.
ISBN 0–470–21709–X
1. Water-supply, Rural. 2. Water quality – Measurement.
3. Drinking water. I. Helmer, Richard, 1943– . II. World Health Organization. III. United Nations Environment Programme.
IV. Title.
TD927.L65 1991
628.1'61 – dc20

Set in 9½/12 Linotron 202 Plantin Roman.
Printed and bound in Great Britain at the Bath Press, Avon.

Contents

Executive summary

Three pilot project areas were selected from three continents (South America, Africa and South-East Asia) for implementing and evaluating WHO (1984/85) Volume 3 guidelines for water surveillance in small communities.

The common objectives of the project were:

1. To develop, test and evaluate water surveillance methods incorporating sanitary inspection and water quality monitoring in rural areas.
2. To develop a surveillance infrastructure at sub-provincial level to ensure that drinking water supplies are kept under continuous public health assessment.
3. To provide a scientific basis for prioritizing remedial action strategies which will protect the consumer from the risk of water-borne disease.

Two of the project countries, Indonesia and Peru, based the project within the existing surveillance infrastructure of the environmental health sections of their health ministries. The third country, Zambia, shared the responsibility between the Department of Water Affairs (DWA) and the Ministry of Health through the framework of the Water Sanitation and Health Education Committee (WASHE). Consequently, the laboratory support in Zambia was provided by the water supply agency, whereas in Indonesia and Peru it was based in the hospitals of the Ministry of Health. It was concluded that financial constraints may preclude countries from developing truly independent surveillance agencies operating in rural areas. Particularly in the early stages of development, the ideals of separation of powers recommended in the WHO Guidelines do not match the realities of monitoring capacity. Some countries will have to depend, at least in part, on the agencies already involved in water supply construction and cooperative action should be encouraged during the development of the surveillance agency.

The surveillance work-force was made up of sanitarians in Indonesia and Peru and of health assistants in Zambia. These people are the key to the success of surveillance and water supply improvement and this in turn depends on their competence, dedication and motivation. It was noted that specialized training was provided and adequate in each of the pilot projects; that motivation depended largely on regular supervision and this was poorly managed; and that dedication was difficult because sanitarians have low status, low remuneration and poor career prospects. This is aggravated by the multiplicity of functions which sanitarians are expected to undertake, apart from water supply surveillance.

Water analysis was focused primarily on detecting bacteriological contamination using faecal indicators. Conventional treatment including chlorination was either not done or ineffective in all three countries, even in piped supplies from surface water sources. Consequently, water quality depended primarily on the quality and protection of the sources. Because the great majority of small community water supplies are contaminated with coliform bacteria it was essential to develop a classification of all water supplies based specifically on the intensity of faecal coliform bacteria. In Peru four grade ranges of contamination were needed: A = 0/100 ml; B = 1–10/100 ml; C = 11–50/100 ml; D = >50/100 ml. Of Peruvian piped systems, 21 per cent were classified grade A, 39 per cent grade B, 17 per cent grade C and 23 per cent grade D. Grade C and D systems were considered to be intermediate and high risk respectively, and requiring urgent attention. Most of the grossly polluted systems were derived from surface-water sources where treatment plants were not installed or not operated.

In Indonesia this classification system was adapted to discriminate between the higher number of more grossly polluted sources, thus: A = 0/100 ml; B = 1–10/100 ml; C = 11–100/100 ml; D = 101–1000/100 ml and E > 1000/100 ml. This classification was applied to all five main categories of point source systems: open dug wells, handpumped dug wells, handpumped shallow and deep tubewells, and rainwater tanks. It was found to be particularly useful in distinguishing the level of contamination in the open and handpumped dug wells of which about 20 per cent of each were in D/E categories and considered to be grossly polluted, and thus the programme focused on protected groundwater sources. In Zambia, although it was reported that over 80 per cent of traditional sources were contaminated with coliforms, very few showed faecal coliform contamination levels greater than 10/100 ml. All borehole and standpipe samples were reported as conforming to WHO guideline values (grade A) and only 7 per cent and 8 per cent of handpumped protected wells and shallow wells respectively were worse than the grades A/B applied in Peru and Indonesia. Thus it was not necessary to extend the grading scheme beyond C for the Zambian project area.

The sanitary inspection methodology for each country evolved from volume 3 of the WHO Guidelines and was quite distinct for each country. That for Peru was the most elaborate because all systems in study were piped community supplies, whereas those for Zambia and Indonesia dealt primarily with point source groundwater systems.

In the early stages of all three projects the data needed for subsequent risk assessment were defective due to ineffective sanitary inspection. It was not possible to relate bacteriological risk to the source of risk. It was therefore essential to reformulate the inspection report forms, retrain the sanitarians and re-evaluate the revised report forms. This process took over 2 years before a working system could be developed which could effectively identify the sanitary risks attributable to a particular source. It was demonstrated that with proper sanitary inspection the principal risks may be regularly identified and remedial action strategies developed.

Cost-effective surveillance procedures were finally developed which ensured that complementary sanitary inspection and quality analysis identified the priority systems requiring remedial action and rehabilitation. Using surveillance results Peru has initiated pilot strategies to tackle the most urgent and high risk treatment systems which have now been completely redesigned to incorporate improved flow control and prefiltration in advance of slow sand filtration. In Indonesia the traditional wells are being protected and replaced by handpumped tubewells, whereas in Zambia the programme of unprotected source improvement has been initiated.

The pilot projects have served as effective models and are being replicated and used as the basis for developing national strategies for implementation of surveillance. It is now vital for the surveillance agencies to institutionalize and formalize these procedures by developing data bases at central level which will provide the rational basis for prioritization of water supply improvement at national level.

Acknowledgements

The authors are indebted to the following for permission to reproduce copyright illustrations:
Mr Jamie Bartram for the cover photograph of a sanitarian, field-testing water in Peru;
Dr Maurice Pardon for Fig 5.6;
All other photographs and original figures have been supplied by the authors.

This book is based on concepts of water supply-related health risk assessment developed by the authors during the course of the Water Decade. These concepts would not have been worth publishing without the practical support provided by several hundred sanitary officers, scientists, engineers, doctors and close colleagues. They have provided the essential information, largely contained in pilot project reports, which have confirmed that a systematic approach to surveillance and risk assessment provides the necessary basis on which to improve and maintain the safety of community water supplies.

This book is therefore first and foremost a tribute to the struggle of environmental health staff everywhere, but in particular to those working in most difficult conditions in the pilot projects in Indonesia, Peru and Zambia. They have generated the data published here.

The health scientists, engineers and colleagues who have been most closely involved in this work met at the Robens Institute in November 1988 to discuss and evaluate the final reports from three rural water quality pilot projects supported by UNEP and WHO. For their many and varied contributions we wish to thank most sincerely: Mr Jamie Bartram, Robens Institute, Guildford; Dr Mauricio Pardon, DelAgua/CEPIS, Lima; Dr Sally Sutton, Chiswick, London; Ms Sri Suyati, Ministry of Health, Jakarta; Dr Hans Utkilen, National Institute of Public Health, Oslo; Dr David Wheeler, Robens Institute, Guildford. We also gratefully acknowledge the financial support of NORAD and ODA.

Barry Lloyd and Richard Helmer
January 1990

List of abbreviations

BST	pumped systems
BTKL	National Sanitary Engineering Laboratory (Indonesia)
CDC and EH	Communicable Diseases Control and Environmental Health (Indonesia)
CEHA	Regional Centre for Environmental Health Activities (Amman)
CEP	Community Education and Participation (Zambia)
CEPIS	Pan-American Centre for Sanitary Engineering and Environmental Sciences (Peru)
CINARA	Centro Inter-Regional de Abaste cimiento y Remocion de Agua (Colombia)
CONCOSAB	National Coordinating Council for Environmental Health (Peru)
COOPOP	'Popular Cooperation' (Peru)
CORDE	Regional Development Corporation (Peru)
DANIDA	Danish Agency for International Development
DIGEMA	Peruvian Ministry of Health, Division of the Environment
DISABAR	Directorate of Basic Rural Sanitation (Peru)
DISAR	Division of Rural Sanitation (Peru)
DITESA	Technical Directorate for Environmental Health (Peru)
DPD	Diethyl paraphenylene-diamine
DWA	Department of Water Affairs (Zambia)
EMINWA	Environmentally Sound Management of Inland Waters
GEMS/WATER	Global Freshwater Quality Monitoring Project (WHO)
GST	gravity systems
GTZ	German Agency for International Technical Cooperation
IDWSSD	International Drinking Water Supply and Sanitation Decade
JAAP	Community drinking water administrative committee (Peru)
lppd	litres per person per day
MF	membrane filter
MPN	multiple tube most probable number method
NCSR	National Council for Scientific Research (Zambia)

NORAD	Norwegian Agency for International Development
ODA	Overseas Development Administration (UK)
PEPAS	Regional Centre for Promotion of Environmental Planning and Applied Studies (Malaysia)
PNAPR	Peruvian National Plan for Rural Water Supply
SENAPA	Peruvian National Water Authority
SIDA	Swedish Agency for International Development
TU	turbidity units
UNDP	United Nations Development Programme
UNEP	United Nations Environment Programme
USEPA	United States Environmental Protection Agency
WASHE	Water Sanitation and Health Education Committee (Zambia)
WHO	World Health Organization

CHAPTER *1*

Introduction

1.1 The basic needs for water and health

The importance of water as a vehicle for the spread of diseases has long been recognized and there are numerous publications and articles concerning health in relation to water supply and sanitation. Most of the diseases which prevail in developing countries, when water supply and sanitation are deficient, are infectious diseases caused by bacteria, amoebae, viruses or various worms. The prevention of any communicable disease requires that the cycle of disease transmission be interrupted. Depending on the prevailing transmission pathways, different interventions in water supply and sanitation are required. Bacterial diarrhoea and epidemics of cholera and typhoid, for example, are often transmitted in drinking water, thus making the quality of drinking water of highest importance. For others domestic hygiene, including sanitary food handling, is of prime importance.

In general, one tends to associate developing countries with microbiological water quality problems and industrialized countries with problems caused by chemical contaminants. In terms of morbidity and mortality rates linked to water quality, this is certainly true and the number of infant deaths attributed to diarrhoeas transmitted by contaminated drinking water is a convincing argument. However, the association of chemicals with the rich countries and of bacteria with the poor countries might prove to be too simple a formula.

The number of water-borne outbreaks of infectious diseases has actually started to increase again in several western industrialized countries, particularly in small community supplies. Thus *Giardia lamblia* and *Cryptosporidium* have become pathogens of public concern in several parts of Europe and North America.

On the other hand, many countries in the developing world are faced with chemical water quality problems which could be of either natural or man-made origin. There are, for example, geologically confined areas in South America, East Africa (Rift Valley) and Central Asia where fluoride in drinking water is at such high levels that mottled teeth and even skeletal fluorosis is occurring. The use of various agro-chemicals, such as nitrogenous fertilizers or chlorinated insecticides and herbicides, is increasingly affecting water resources which are vital for community water supply.

In either case, it is the rural populations which are suffering the most from such microbial or chemical contamination of drinking water. Their numbers are overwhelmingly high and increasing at a rate which makes it extremely difficult for water supply and sanitation programmes to achieve significant improvements over and above demographic developments. The needs of the rural areas in the developing countries are shown by the fact that in 1980 less than one-third of the rural population was served with satisfactory drinking water supply schemes as compared with about three-quarters of the people living in towns (WHO 1988d).

Much could be achieved in terms of improved health and reduced infant mortality if water supply services were up to standard. Although it is quite obvious that pathogen-contaminated drinking water is a prime source of infection, it is equally true that insufficient availability of water hampers people's efforts to practise good personal and domestic hygiene. The inevitable consequence is high diarrhoeal and skin disease incidence. Four major indicators are proposed in this book to assess the adequacy of community water supplies:

Coverage: the proportion of the population served and the proximity of the water point to the place of use;
Continuity: the reliability of supply throughout the day and throughout the year;
Quality: bacterial and chemical quality of the water;
Quantity: the quantity of water available per user per day.

The question remains as to what extent improvements in water quantity/quality and sanitation facilities would result in an improved health status of the rural population. Although it is recognized that quantification of such cause–effect relations is rather difficult, there is no doubt about the principal health benefits. In Table 1.1 an attempt is being made to attribute relative health improvements to different sanitary engineering methods.

Table 1.1 Percentage reductions in diarrhoeal morbidity rates attributed to water supply or excreta disposal improvements (Esrey *et al.* 1985)

		Percentage reduction	
Type of intervention	Number of results*	Median	Range
All interventions	53	22	0–100
Improvements in water quality	9	16	0–90
Improvements in water availability	17	25	0–100
Improvements in water quality and availability	8	37	0–82
Improvements in excreta disposal	10	22	0–48

* There are 53 results in total but only 44 attributed to specific interventions. The remaining 9 results are for other interventions or combinations of interventions having less than 3 results, and include interventions in fly control and health education together with water supply or excreta disposal.

Since the results from the studies evaluated in Table 1.1 vary over a wide range, it may be equally instructive to look at a specific country example. In Costa Rica, which is the case presented in Fig. 1.1, the effects in terms of reduced diarrhoeal mortality rates became tangible after each intervention programme in the water and sanitation programme. In particular, the decisive finishing-off with water-borne diseases by treating the water for better quality is noteworthy (see programme A in Fig. 1.1).

The Costa Rica example also highlights the conclusion that in all instances basic requirements have to be met, not only for the microbial and chemical quality of water but also for the availability of the necessary quantities of water, the convenience of the water source and the reliability of the supply. In order to achieve and maintain a pathogen-free human environment, it is mandatory that all potential routes of exposure be barred off and the cycle of disease transmission be effectively interrupted. This was impressively demonstrated for the first time by John Snow when he removed the pump handle from the Broad Street well pump in London in 1854 and thereby stopped the source of vibrio cholerae transmission from sewage-polluted Thames water. It has become clear today that several concurrent interventions are mandatory to achieve this effect, including water supply, excreta disposal, personal and community hygiene and food safety. The health benefits from these and other concurrent primary health care measures, such as improved nutrition, are real, although they would be difficult to quantify or attribute to single interventions.

Figure 1.2 conceptualizes the direct and indirect health benefits and indicates the links with water quantity and quality issues.

1.2 The international response

It was already considered at the United Nations Conference in Mar del Plata, 1977, that 'all peoples, whatever their stage of development and their social and economic conditions, have the right to have access to drinking water in QUANTITIES and of a QUALITY equal to their basic needs'.

In pursuing the fulfilment of this right, quantitative and qualitative aspects of supplying safe and adequate drinking water have to be equally taken into consideration:

(a) *safe water*, i.e. water free from chemical substances and micro-organisms in concentrations which could cause illness in any form; and
(b) *adequate water supplies*, i.e. providing safe water in quantities sufficient for drinking, culinary, domestic and other household purposes including the personal hygiene of members of the household. It should also provide sufficient quantities on a reliable, year-round basis near to the household where the water is to be used.

These basic needs were also recognized and endorsed by the WHO/UNICEF International Conference on Primary Health Care in Alma Ata, 1978, which included the provision of adequate supplies of safe drinking water and basic

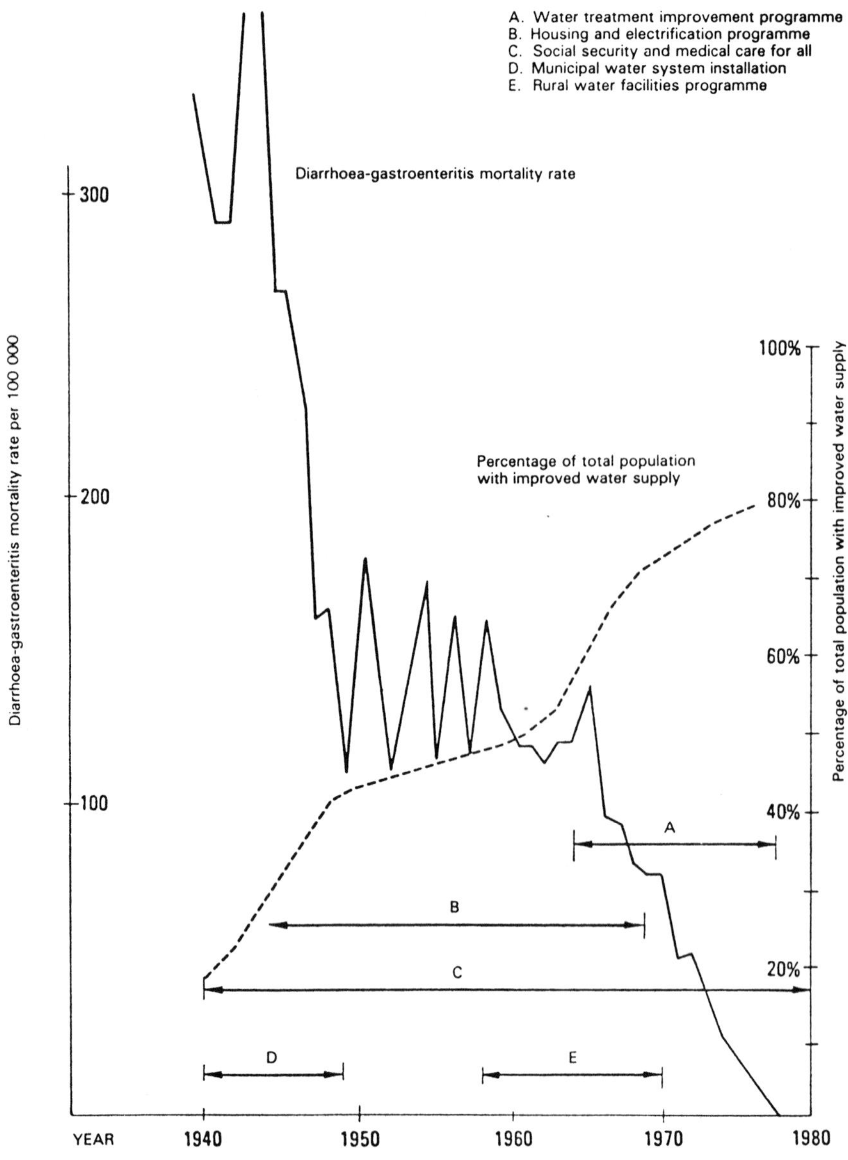

FIG. 1.1 Diarrhoea–gastroenteritis mortality rates versus time, and percentage of total population with improved water supply versus time, Costa Rica (McJunkin 1982). (A) water treatment improvement programme; (B) housing and electrification programme; (C) social security and medical care for all; (D) municipal water system installation; (E) rural water facilities programme.

sanitation as one of its essential strategy elements. Within the subsequently declared International Drinking Water Supply and Sanitation Decade (IDWSSD), the WHOs strategy on national institutional development emerged in the form of two complementary objectives:

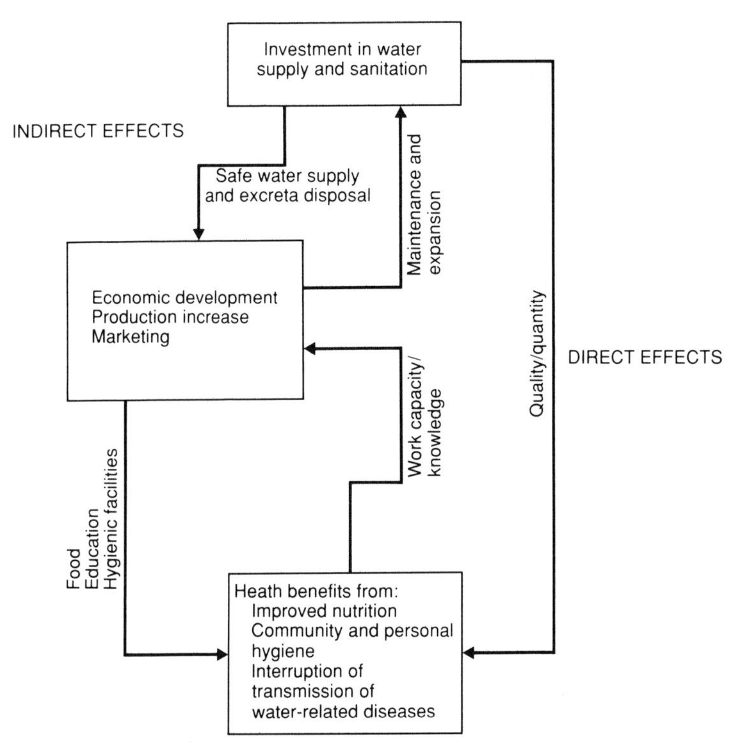

FIG. 1.2 Direct and indirect effects of water supply and sanitation on health: a conceptual framework (Cvjetanovic 1986).

1. To support governments in strengthening health agencies in their roles of monitoring the health impacts of drinking water supply and sanitation programmes, promoting improvements, and coordinating these programmes with other components of primary health care; and
2. To cooperate with governments in the establishment of appropriate quality standards for drinking water, in the organization of national drinking water quality surveillance programmes, and the protection of drinking water sources.

Surveillance implies that regular and routine surveys be undertaken to provide for a systematic series of observations concerning water quality in order to ensure that quality standards are consistently being achieved. An effective surveillance programme depends on the existence of national regulatory standards of water quality and codes of practice. These, in turn, depend on appropriate national legislation and the establishment of a competent surveillance unit or agency within government.

During the first years of the IDWSSD, there was little information available on the extent of water surveillance programmes in place in developing countries. The few data at hand indicated that regular surveillance of drinking water quality in the rural areas is rare in any developing country. Whereas urban waterworks usually conduct basic bacteriological and chemical monitoring, the small communities, particularly if they are dispersed over vast rural areas, do not have the facilities and are seldom covered by routine services.

The reason for this situation is quite clear and has been reported on many occasions by health officials from developing countries. They often do not have the resources in terms of manpower, laboratory facilities and logistics to permit the establishment of continuous and systematic programmes for the surveillance of drinking water quality, nor, unfortunately, is a high priority attached to this activity. Top priority was assigned early in the IDWSSD to coverage and installation of new water supply facilities – less attention was given to the adequate operation and maintenance of existing systems, to the detection of inadequacies of existing systems and to the correction of deficiencies. A general attitude prevailed that the supply of adequate quantities of water must take precedence over activities to ensure safe water quality. This does not suggest in any way that health officials did not recognize the benefits to health that can accrue from effective water quality surveillance. It did suggest, however, that those health benefits were not fully appreciated outside the health community.

As concerns the progress achieved during the IDWSSD programme in terms of water availability and coverage with relevant supply services, an interim assessment was undertaken by WHO at the end of 1988, i.e. after 8 years of the IDWSSD programme (WHO 1988d). For rural populations, it was estimated that an additional 310 million persons received access to an adequate and safe water supply during that period. This, however, still leaves approximately 915 million unserved. In other words, for every one person provided since the start of the IDWSSD, there remain about three without services.

Nevertheless, there was a rise in the overall service coverage from 31 to 46 per cent and the disparity between urban and rural residents has been reduced accordingly. Table 1.2 demonstrates the great achievements, particularly in South-East Asia, which means that about half of all rural people will be adequately served with safe drinking water by the end of the IDWSSD.

Experience during the 1980s has also underlined the dangers of regarding water supply and sanitation simply as a question of constructing, operating and maintaining facilities and developing the necessary institutional and support infrastructures. Relevant programmes are increasingly being implemented within the framework of a broad water resource management approach. Although the quantities needed for domestic supply are usually small in comparison to irrigation water and livestock watering demands, the quality requirements for human uses are rather stringent.

An evaluation of the full impact of the IDWSSD on health would have to take into account reductions in incidence of typhoid, dysentery, poliomyelitis, hepatitis, schistosomiasis, filariasis, leptospirosis, trachoma, dracunculiasis, scabies, amoebiasis, and other water-borne or water-related diseases and those with vectors depending on aquatic environment. Also, the mineral content of drinking water affects such conditions as the occurrence of methaemoglobinaemia, dental caries or fluorosis. The indirect benefits of better access to satisfac-

Table 1.2 Rural water supply: population covered in developing countries (excluding China) (WHO 1988)

WHO Region	1980 (millions) (%)	1988 (millions) (%)	1990 Estimate (millions) (%)	1990 Target (millions) (%)
African	61,585 (22)	88,573 (26)	95,320 (27)	183,218 (52)
Americas	51,243 (41)	59,727 (47)	61,847 (49)	64,354 (51)
South-East Asia	253,089 (31)	513,249 (46)	577,789 (62)	610,243 (65)
Eastern Mediterranean	53,812 (31)	56,336 (28)	56,967 (27)	139,434 (67)
Western Pacific	46,353 (41)	62,003 (50)	65,916 (52)	96,048 (76)
Global	468,083 (31)	779,888 (46)	851,942 (49)	1,093,298 (62)

tory water sources include improved nutrition and general health status among children (see Fig. 1.2).

Two diseases can be used for a preliminary assessment of the IDWSSDs impact: diarrhoeal disease and dracunculiasis, which is the only disease that can be totally eliminated from a community through safe water supply.

The United Nations Water Conference, which initiated the IDWSSD, clearly emphasized that the main concern was for health improvement and prevention of diarrhoeal diseases and infant mortality and morbidity in developing countries. In the first 8 years of the IDWSSD, a total of some 310 million more people in rural and 225 million more in urban areas in developing countries gained access to water supply systems classified as adequate and safe. Approximately one-third, i.e. 180 million, are children under the age of 5 years – the most vulnerable age-group. Recent studies indicate that safe water can reduce diarrhoeal disease incidence by approximately 30 per cent, as shown in Table 1.1. Thus, since the start of the IDWSSD, 60 million out of about 180 million episodes of diarrhoeal disease in children under 5 can be estimated to have been avoided (WHO 1988d).

The need for a safe water supply in order to eradicate dracunculiasis is undisputed. Unlike diarrhoeal disease, which is a scourge in all developing countries, dracunculiasis only seriously affects 23 countries, 20 of which are on the African continent. Ten countries, including Ghana, India, Nigeria and Pakistan have reported national plans against dracunculiasis. A major achievement, for example, has been the drop in cases reported by India from 30,440 to 14,296 since 1985. This was largely attributable to the national eradication programme launched at the start of the IDWSSD, which directed resources for rural water supply to villages affected by the disease (WHO 1988d).

These are just a few examples which illustrate the close interlinkage between drinking water quantity/quality and human health. Although the present publication concentrates more on water quality aspects, it should be kept in mind that safeguarding drinking water quality is only one component in the strategy to eliminate water-related diseases.

1.3 The WHO Guidelines

The pressing need to provide safe drinking water was the motive for preparing during the early years of the IDWSSD new WHO Guidelines for Drinking Water Quality (1984/85), which replaced the earlier International and European Standards. The emphasis of the Guidelines was placed first and foremost on the microbiological safety of drinking water supplies as more than half of the world's population is still exposed to water that is not free from pathogenic organisms, resulting in infectious diseases that ultimately lead to increased mortality rates in the population. It is quite clear that in those areas guidelines and programmes for assuring the chemical safety or the organoleptic qualities of drinking water are of much lesser urgency, except in locations where industrial effluents or agricultural chemicals are seriously endangering the water supply.

The last edition of the WHO *International Standards for Drinking Water* was issued in 1971 and that of the *European Standards* in 1970. These standards were reviewed, revised and combined with substantive support from DANIDA and issued in 1984 under the series title of the WHO Guidelines for Drinking Water Quality. The preparation of the new Guidelines took over 3 years and involved the active participation of nearly 30 national institutes, literally hundreds of scientists and meetings of 10 task groups.

Possibly the best illustration of the change which has taken place in the basic approach used is the change of the title of the publication itself, i.e. from 'Standards' to 'Guidelines'. This change is intended to reflect more accurately the advisory nature of WHO recommendations so as not to confuse them with legal standards which are the responsibility of the appropriate authorities in each country. The new Guidelines clearly recognize the desirability of adopting a risk–benefit approach, qualitative or quantitative, to national standards and regulations. The establishment of drinking water quality standards must follow a very careful process in which the health risk is considered alongside other factors such as technological and economic feasibility. The establishment of standards without considering the practical measures which will need to be taken with respect to either finding new sources of water supply, instituting certain types of treatment and in providing for adequate surveillance and enforcement, will not yield the desired results. In the Guidelines, the need for careful consideration of the standard-setting process, including follow-up activities, is very much emphasized.

In light of the varying priorities with respect to drinking water quality around the globe, it was considered necessary to issue the new Guidelines in three volumes. Each of these was intended to serve a different purpose and, to a certain extent, they were also directed towards different audiences.

The three volumes of the new WHO Guidelines (WHO 1984/85) contain, in brief, the following:

Volume 1 – *Recommendations* presents the recommended guideline values, together with essential information required to understand the basis for the recommended guideline values as well as information on monitoring requirements and, where possible, suggestions regarding remedial measures to ensure compliance with the guideline values. It provides guidelines with respect to microbiological, biological, chemical, organoleptic and radiological quality of drinking water.

Volume 2 – *Health Criteria and Other Supporting Information* sets out the health criteria for those drinking water pollutants and other constituents which were examined, as well as providing information regarding detection of contaminants in water and measures for their control. It contains a review of toxicological, epidemiological and clinical evidence which was available and used in deriving the recommended guideline values.

Volume 3 – *Drinking Water Quality Control in Small Community Supplies* deals specifically with the problem of small communities, predominantly those in rural areas of developing countries. It contains information on techniques for the assessment and control of contamination of such supplies. It covers simple methods for sampling and analysis, sanitary surveys and other means of investigating and protecting drinking water supplies in these areas. In addition it summarizes the lines of communication within the surveillance agency and with the water supply agency from local to national level (Fig. 1.3).

It is the objective of providing safe drinking water, with the emphasis on 'safe' that makes the Guidelines an integral element of the IDWSSD. Thus the three volumes of the Guidelines provide not only an important tool to secure safe supplies, but also to set the yardstick for measuring progress and achievements towards this IDWSSD goal. Countries still suffering from water-borne infections should find useful support from the microbiological and biological parts of the guidelines, and from the advice on their application. Volume 3 is specifically intended to bridge the gap between the mere publication of guideline values and their actual compliance often under adverse hydrological and socio-economic conditions (Helmer and Ozolins 1987).

1.4 Focus on the rural areas

For many developing countries, the WHO Guidelines should be of direct use in developing and implementing their national IDWSSD plans for water quality control. Bearing in mind the particularly serious health situation in the rural areas of most developing countries, the need for a special effort towards the safeguarding of their water supplies was recognized and a special guideline was prepared to deal with this situation.

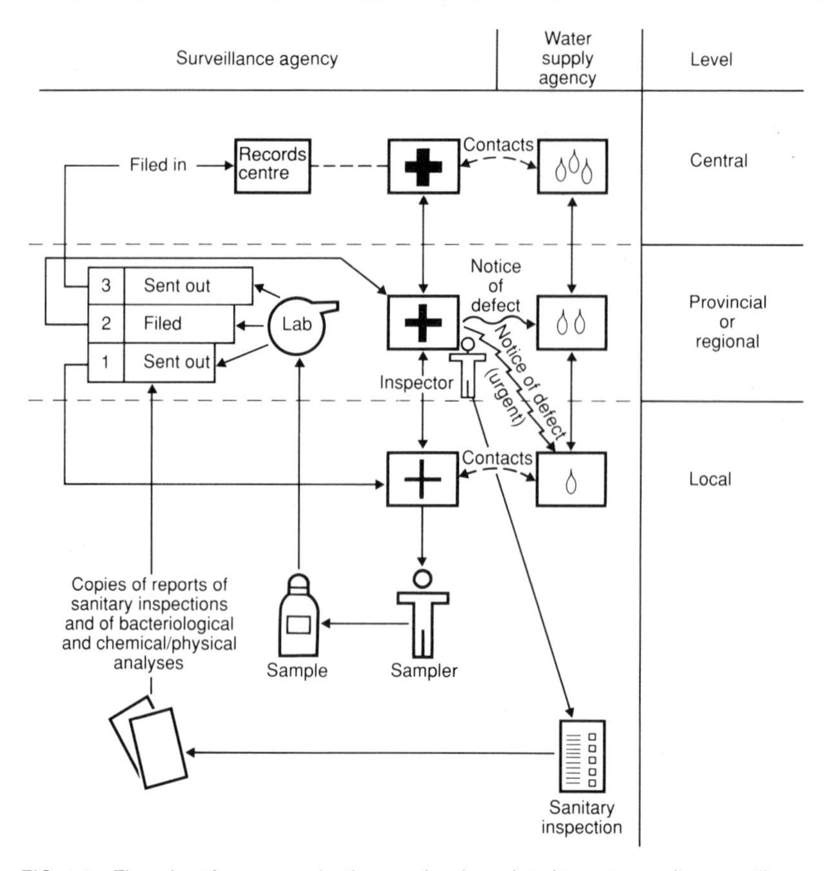

FIG. 1.3 Flow chart for communications and action related to water quality surveillance.

Securing safe drinking water for small populations, often scattered over large geographical areas, poses tremendous logistic problems for which standards or guideline values alone can at best provide a mere reference point. What is the meaning of a comprehensive list of national standards or guidelines when there are just a few district hospital laboratories available with limited capacity and interest for analysing water samples? There, the concern for water quality can be brought back to its essentials, namely the microbiological safeguarding of the water supplies. Although the recommended guideline values for bacterial quality of drinking water ought to be aimed at as a goal, the approaches and means to safeguard the drinking water supply will need to be tailored to the situation in each country. Rural installations are shown in Figs 1.4 and 1.5.

As part of the overall development of the new WHO Guidelines a DANIDA-supported meeting was convened in late 1982 in Bangkok to deal specifically with drinking water quality surveillance for small communities in rural areas. Contrary to the water quality surveillance schemes which have been designed for large urban water supplies, little has been achieved so far in the rural areas

FIG. 1.4 Rainwater catchment in Indonesia.

FIG. 1.5 Handpump installation in West Africa.

and their small communities. As a first step, methods were to be identified that allow for a cost-effective application of the WHO Guidelines in sparsely populated areas. This required not only that simple monitoring and control methods be made available, but also that the responsible health authorities at national, state and district level be ready to adopt and implement them.

Advice on what needs to be done was put together in volume 3 of the WHO Guidelines. Taking into account the often severe constraints and likely difficulties communities are facing in rural areas, main emphasis was placed first and foremost on the microbiological safety of the drinking water supplies. The information provided should allow for the selection of methods and techniques in accordance with local conditions, and likely limitations of the available resources and manpower. Where possible, simple well-tried techniques were advocated.

Following expert advice received at the Bangkok meeting and drawing on field experience gained in places as far apart as Papua New Guinea, Thailand and Botswana, the WHO Guidelines became the basis for a new WHO strategy for drinking water control in rural areas. Major elements were considered to be the following:

1. Careful planning for surveillance and control, including the design of a workable organizational structure, the assessment of local conditions and the proper handling and use of information. Special attention is needed for those communities where no surveillance is yet in place and where future possibilities seem to be limited.
2. Sanitary inspection is advocated as a crucial part of any surveillance programme which needs to be carried out by trained people at the community or regional level. In many poverty-stricken areas this activity may be the only feasible form of surveillance for the time being.
3. Field and laboratory methods should be made available for standard microbiological analysis of drinking water, including the membrane filtration and the multiple tube methods for both total and faecal coliforms. Attention has to be given to arrangements for transport, where necessary, of samples to the nearest laboratory.
4. In the case of chlorinated water supplies, the routine checking of residual chlorine is regarded as essential and field methods should be provided. This determination is amenable to application at the community level with a minimum of training.
5. Immediate remedial measures as well as long-term and short-term preventive action are considered absolutely essential for achieving the desired control of drinking water quality. Only if sanitary deficiencies identified during surveillance are remedied will the effort put into the programme render its full benefits.
6. Community education and involvement are other mandatory components upon which an effective surveillance programme must be built. The active participation of the members of the community is considered an essential prerequisite in safeguarding drinking water quality, particularly in remote areas with small communities that are widely scattered. Much of the local health education is to be implemented within the framework of primary health care.

Bearing in mind the complexity of the infrastructure required, surveillance responsibilities ought to be shared and coordinated between the water supply agencies and the health authorities. The latter, in particular, have the overall responsibility for ensuring that all drinking water under their jurisdiction is free from health hazards. All too often, however, the health authorities have neither the necessary programmes nor the qualified staff to implement them.

Fulfilling this function involves both sanitary inspections and the sampling and analysis of public supplies. Ideally, this involves legislation supported by regulatory standards and codes of practice, trained staff, laboratory installations, coverage by regular health inspection services, educated community health workers, etc. However, a pragmatic programme may have to be started with only a fraction of these requirements at the initial stage.

1.5 Project aims and objectives

Three broad fields of action appear to be of crucial importance. The first is to heighten awareness among government officials and the public of the need for water quality surveillance and control. The second is to cooperate with governments in the establishment of surveillance and all that such establishment implies – training, laboratory supplies, reporting mechanisms and the initiation of corrective measures whenever necessary. Finally, the third necessity concerns the development and dissemination of approaches and techniques appropriate for the usually rather limited infrastructure of rural areas. The present book has been prepared to demonstrate how activities in these areas have been pursued simultaneously in three countries to meet the needs for safe water supplies in small communities remote from the national centres of technology and resources.

Preparation and discussion of volume 3 of the WHO Guidelines, *Drinking Water Quality Control in Small Community Supplies* (WHO 1984/85) made it immediately clear that good advice and production of strategy documents alone are not enough to get the water and health sectors moving together towards comprehensive rural area programmes. Examples had to be given and the feasibility of the proposed approaches demonstrated.

The WHO and the United Nations Environment Programme (UNEP) as the two leading UN agencies in the area of water quality joined forces in 1984 to launch a project in support of application and verification of the WHO Guidelines. The project's long-term objective was 'to support the aims of the International Drinking Water Supply and Sanitation Decade through the promotion of effective control mechanisms for safeguarding drinking water quality' (Helmer 1989). The immediate aims were:

1. to apply and field test methods for the surveillance of drinking water quality under conditions of rural areas;
2. to establish demonstration areas for the effective control of drinking water quality in small community supplies;
3. to train technical staff of national and local authorities responsible for drinking water quality;

4. to field test and, where necessary, improve the WHO Guidelines for the surveillance and control of drinking water quality in small community supplies.

Through the project, simple methods were to become available at the community level, and health authorities were to receive assistance in the design and implementation of surveillance services. Much experience could be drawn at that time from ongoing country projects which received bilateral support through, e.g. the Norwegian Agency for International Development (NORAD) in Tanzania, German Technical Cooperation (GTZ) in Thailand and Swedish International Development Agency (SIDA) in Botswana. At the same time innovative approaches towards the design of the necessary equipment, field kits, etc. were taken by research and development teams in different places around the world. Simplified test equipment for bacteriological, physical and chemical analysis of drinking water were developed simultaneously in countries such as Peru, India, China and the United Kingdom. Simple processes for water chlorination and for defluoridation were emerging in Argentina and India. In all, the time was right to bring the many developments together and apply them within the same pilot areas. Thus, their reliable functioning under different, more or less difficult, conditions could be demonstrated.

Success or failure of the pilot activities – and this was obvious from the inception of the project – would largely be determined by the human factor, i.e. the competence and dedication of all participants. Therefore the responsible health authorities and water agencies were to be brought together in working sessions during which approaches and methodology were to be introduced, and technical as well as manpower requirements for the establishment of the pilot surveillance schemes discussed. In each case the health requirements, specific constraints, socio-economic conditions and prevailing cultural aspects were to be determined. Thus a common basis for active participation of staff at all levels was to be generated. The incorporation of village health workers, sanitary inspectors, public health officers, district hospitals, etc. in varying geographic, demographic and administrative settings was one of the prime challenges in developing each pilot area programme.

Preparatory discussions with several interested countries and external support agencies led in 1985 to the designation of pilot projects in three countries, each being located in a different continent. They were: Gunung Kidul regency in Indonesia, Health region XIII in Peru, Mongu district in Western Zambia.

The objectives and designs of each pilot project were somewhat different in as much as local and national requirements had to be taken into account. Additional technical and financial support was granted by bilateral agencies, i.e. the UK Overseas Development Administration (ODA) in the case of Peru, and NORAD in the case of Zambia. As seen in Table 1.3, there are similarities between the three surveillance programmes in the specific objectives of their pilot stage.

The training of personnel and development of methodology for further

Table 1.3 Specific objectives of the water quality surveillance demonstration projects (Pardon 1987)

Objectives	Indonesia	Peru	Zambia
Formulation of technical standards		*	
Training of personnel	*	*	*
Implementation of laboratories		*	
Promoting or reporting of information	*	*	
Development of methodology for further replication on a national scale	*	*	*
Identification of health impact			*
Promotion and development of corrective activities	*	*	*

replication were the most conscientiously and closely followed objectives. In the case of the development of methodology, this objective was achieved during the course of the projects.

In the formulation of standards, the three programmes were still cautious, using the WHO Guidelines as a basis but without attempting to develop them further or to enforce them. However, in the Indonesian and the Peruvian programmes, the degree of faecal contamination of water supplies was categorized, thus providing a basis for prioritizing for corrective action/intervention.

The development of health impact studies requires a complicated methodology and was attempted as part of the collaborative activities of the Zambian programme. The Zambian project reports do not present evidence, however, that this objective has actually been achieved during the course of the project.

The promotion and development of corrective activities requires two main components: (i) the existence and practical experience of viable strategies, and (ii) an institutional structure prepared to implement it. Here, the Indonesian and Peruvian programmes report initiatives being taken on the first aspect, but there is no mention in any of the reports of the second aspect being dealt with effectively.

The activities in each pilot area were supported mainly through the following measures:

1. Equipment provision for bacteriological tests (sampling boxes, membrane filtration units, incubators, field kits, etc.).
2. On-the-job training of technicians and sanitarians involved in sanitary inspections, sampling/analysis and advisory services on remedial/preventive measures.
3. Development of local language documentation and audio-visual aids for related health education and specific technical instructions.
4. Teaching of trainers in health education through courses provided at the demonstration sites.
5. Promotion of community education and involvement by local trainers throughout each demonstration area.

It was the hope of the project agencies that the pilot areas would, by the end of the pilot phase in 1988, develop into continuous local surveillance activities for the benefit of the communities served in those areas. It was also envisaged, however, that the experience and results of the pilot exercises would serve as valuable examples from which better and more effective guidance could be drawn. In particular, a comprehensive and field-tested package for organizational and technical alternatives for drinking water quality control in rural areas should emerge as a major project output, valuable for many other countries in comparable situations.

It is with this latter purpose in mind that the present publication has been prepared. In the following chapters the background, implementation and experience of the three pilot areas are described in greater detail. In Chapter 7 the surveillance results obtained in each area are discussed and their health significance evaluated. Based upon an application of these results, but also drawing on information from other similar country projects, conclusions are reached concerning the application of the WHO Guidelines in typical rural situations with many small-community supplies, often scattered over vast areas.

Thus, volume 3 of the WHO Guidelines, together with the present book, provide the basis for the current WHO strategy for technical cooperation in this sector. As described in Chapter 9 there are two lines which such cooperation could follow. One is the direct build-in of water quality control components into rural area development programmes and rural water supply projects, for which guidance and methodology have been made available. The other is the broad-scale dissemination and further development of simple and reliable techniques supported by an effective infrastructure for their application.

The momentum gained in this subject area can best be demonstrated by the fact that the pilot project has entered a second phase in 1988 with two more pilot areas, one in East Africa and one in the South Pacific. In addition, more and more developing countries and their external support agencies are initiating national IDWSSD activities with the aim of controlling drinking water quality countrywide. The three pilot areas established in phase one of the project became valuable demonstration sites whereby the training of health and water agency staff of neighbouring countries contributed to the transfer of experience from one developing country to the next.

CHAPTER 2

The Three Pilot Projects

2.1 Choice of project areas

The three countries selected to evaluate the usefulness of volume 3 of the WHO Guidelines had obvious and profoundly differing ethnic, political, social and physical characteristics. All three countries lie in the pantropical zone on or south of the equator, but Indonesia is a multi-island nation with little climatic variation, while Zambia is landlocked and Peru, due to the combination of its Pacific edge and continuous mountain chain separating the interior, experiences extremes of climate. Population growth in these countries is in the range 2–3 per cent, but whereas in Peru almost all of this growth is in the urban sector in Indonesia it is the reverse. Greater than 80 per cent of the population in Indonesia is classified as rural, and whereas 60–70 per cent in Zambia is considered to be rural and in Peru less than 30 per cent is classified rural, these proportions are representative of each of the three continental regions. Density varies greatly. A summary comparison is provided in Table 2.1.

Table 2.1 Basic population data in the countries of the pilot projects

	Indonesia	Peru	Zambia
Country			
Total population (millions)	170	20	7
Project region	Yogyakarta	Central	Mongu
Area (km^2)	1,439	70,487	10,075
Total population (thousand)	690	1,352	120
Rural population (thousand)	560	272	80
Rural population (%)	>80	20	67
Rural community classification criterion	<5,000	<2,000	<2,000

In each country pilot project areas were selected to demonstrate and initiate the planning and implementation of water surveillance as a model for wider and eventually national replication. In Indonesia the district (regency) of Gunung Kidul on the most populated island, Java, was selected for the pilot project because it was an area where all types of water supply facilities have been constructed. It also reflected the national strategy which emphasizes the development of wells for groundwater use at neighbourhood and private family level. Likewise in Zambia the selection of Mongu district in the Western Province reflects the emphasis on drawing on groundwater sources. In Peru there are three distinct geographical and climatic zones: coastal desert with almost no rainfall, high sierra with seasonal rainfall and copious spring and stream sources, and tropical rain forest with year-round rainfall. The central sierra was selected for the pilot project as having the greatest variety of developed community water resources.

Table 2.2 Rural water supply facilities in the pilot project areas

	Indonesia (Gunung Kidul)	Peru (Junin)	Zambia (Mongu)
Water treatment plants	0	30	0
Other piped supplies	13	440	14
Surface-water sources	91	>30	Few
Spring protection	45	Included in other piped	Few, not considered
Shallow handpump	459	Few, not considered	>40
Deep handpump	808	Few, not considered	?
Traditional dug well	9,204	Few, not considered	>1,100
Rainwater tanks	11,027	Not recorded	Not recorded

Thus population and water resource distribution were the two most important factors which dictated the national policies and priorities for the type of rural water supply development and the differences are fundamental, as shown in Table 2.2. The commonest traditional rural water sources in Indonesia are dug wells, in Zambia they are wells and water-holes and in the Peruvian highlands springs and streams. This is related to distinctive population distributions which are more dispersed and homogeneous in Indonesia and Zambia, in contrast with the typically nuclear communities and ribbon development found in the highland valleys of Peru (Fig. 2.1). It will be seen later that these characteristics have largely dictated the surveillance strategy developed in each country.

2.2 The Indonesian project

In April 1975 the Government of Indonesia established national standards for drinking water (Government of Indonesia 1975). At about the same time an undated document titled 'Control system of drinking water quality and water

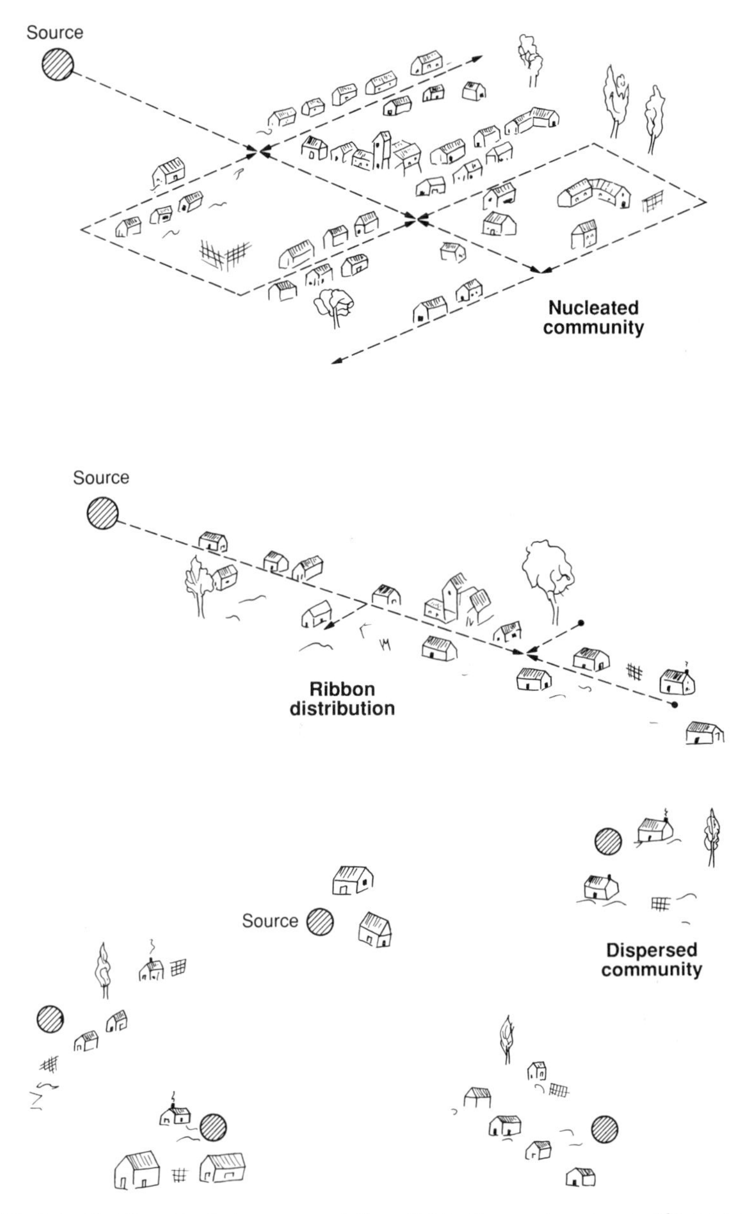

FIG. 2.1 Relationship between population distribution, water source (⊘) and piped supplies (– – –) in rural areas.

pollution', was produced by the Directorate General of Medical Care. As a result a national system of laboratories was established at provincial level for drinking water surveillance focused on the main urban areas. The Director-General of Communicable Disease Control and Environmental Health (CDC and EH) assumed responsibility for drinking water surveillance in both urban and rural areas following a National Drinking Water Surveillance Workshop held in Bali in 1981.

Proposals for a pilot project (WHO Project No. INO/SWC/001) to extend water surveillance to the rural areas were submitted with the support of a WHO consultancy in November 1984. In February 1985 volume 3 of the WHO Guidelines, *Drinking Water Quality Control in Small Community Supplies* (WHO 1984/85) was launched at a 1-week national training workshop in Yogyakarta in Java. A pilot demonstration project to implement and test these guidelines was initiated in the Gunung Kidul district of Yogyakarta province in April 1985 (Lloyd and Suyati 1989).

Yogyakarta province was selected as the target area for the pilot project mainly because the National Sanitary Engineering Laboratory (BTKL) is located in the capital of this province. BTKL is equipped to carry out a full range of analytical functions for water and waste water; it provides a reference service at national level for microbiological, chemical and physical analysis. The chemical laboratory is set up to handle both inorganic and organic analysis including pesticides. BTKL provides training expertise and was thus a convenient centre for training sub-district provincial staff. It was also intended that BTKL should provide logistic and supervisory support to whichever regency/ district (*kabupatan*) was selected for implementation.

Java is the most populated (with more than 120 million inhabitants) of all the islands which make up Indonesia (see Table 2.1) and is divided into 4 provinces and 105 administrative districts. Yogyakarta is the smallest province in Java and is split into 5 districts. The urban municipality (*kotamadya*) of Yogyakarta is thus separated from the 4 rural districts (*kabupatans*). Gunung Kidul is the largest of the 4 rural districts of Yogyakarta province, making up about 47 per cent of the whole province 1439 km^2 in area. The district of Gunung Kidul is divided into 13 sub-districts, 144 villages and 1402 sub-villages (see Fig. 2.2). The total rural population of Gunung Kidul comprises 560,000 inhabitants.

Gunung Kidul was selected for the pilot project because of the variety of water sources which have been developed there. Although the commonest water facilities are neighbourhood dug wells and handpumps there are also many thousands of rainwater collection systems, a few protected springs, artesian wells, surface-water sources and associated piped supplies.

The capital of Gunung Kidul district, Wonosari, is 1.5 hours by road south-east from Yogyakarta and thus readily accessible for supervisory and logistic purposes. Wonosari Hospital therefore became the laboratory base for the pilot project and the local project coordinator. None the less the authority legally responsible for managing surveillance activities and follow-up is the district health officer in Wonosari.

Geographically, the province of Yogyakarta occupies a central position on the southern coast of Java where rainfall is markedly seasonal. The dry season normally lasts for 6 months, from December to July, but in 1987 it continued through September and October with disastrous consequences in those areas mainly dependent on rainwater catchment. Annual rainfall averages 2000 mm

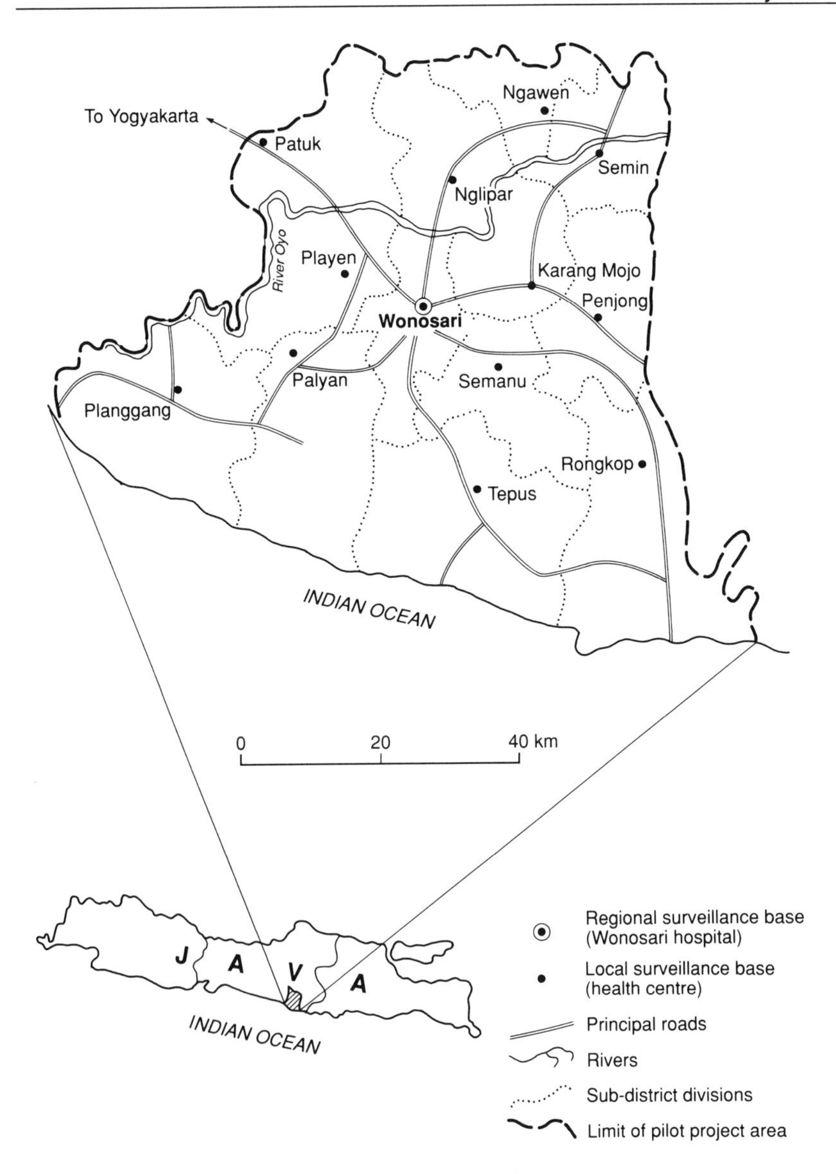

FIG. 2.2 Gunung Kidul district pilot project in Java.

with pronounced seasonal variation. Average temperature is 25 °C with little seasonal variation. The great majority of the province is intensely cultivated with staple crops and the terrain rises from sea-level to over 700 m above sea-level with a number of higher volcanic peaks. The northern part of the district of Gunung Kidul includes four sub-districts at an elevation of 200–700 m above sea-level, mainly north of the Oyo river. In this area there are some springs and surface-water sources.

The central area of Gunung Kidul district includes five sub-districts at an elevation of 150–300 m surrounded by mountains, the so-called Wonosari valley. Here the rivers are seasonal and thus dry for part of the year. The two aquifers, which are located at depths of 8–40 m and 60–100 m, are consequently vital sources of water supply.

The southern part of the district includes four sub-districts with an elevation from sea-level to 300 m. Here there are no surface rivers, due to the high porosity of the dominant geological stratum, weathered limestone, and the main water sources are therefore shallow aquifers, underground streams and rain-water collection.

2.3 The Peruvian project

The legal, political and administrative basis of the programme derives from Peruvian water legislation (Ley de Aguas, 13997, Codigo Sanitaria DL 17405 and Codigo Sanitario de Alimentos DL 102/03). These laws set out the needs for protecting drinking water sources, disinfection and treatment of supplies and water sampling. The present project is attempting to fulfil the spirit of the law by bringing into effect the WHO Guidelines with particular reference to volume 3 for rural supplies.

A preparatory rural water quality project was initiated by the Directorate for the Environment of the Ministry of Health in mid-1984 before volume 3 of the WHO Guidelines became available in Spanish. The DelAgua group of the Robens Institute, based at the Pan American Centre for Sanitary Engineering and Environmental Science, CEPIS/PAHO/WHO Lima, was invited to organize training and supply water testing equipment and consumables for a water surveillance course for sanitary and laboratory technicians. This was done with administrative support from the Ministry Division of Basic Rural Sanitation. The preliminary training course was launched in the Hospital Carrion in the capital of the Department of Junin, Huancayo. This was executed in July 1984 and followed by pilot surveillance activities limited to 60 villages with piped supplies (Lloyd *et al.* 1989).

As a result of the preparatory activity, PAHO/WHO promoted and supported a bilateral aid agreement between the ODA of the Government of the UK and the Government of Peru, with the aims and objectives of enabling the Peruvian Ministry of Health, Division of the Environment (DIGEMA), to initiate the following activities aimed at compliance with existing water quality legislation:

(a) formulate and revise technical standards for the surveillance and control of drinking water quality;
(b) train and evaluate the work of sanitary and laboratory technicians involved in water surveillance;
(c) promote and develop water quality control at the operator level in the health areas;
(d) promote and secure the implementation of water surveillance laboratories;

(e) promote and develop water quality data reporting at regional and national level;
(f) develop regional engineering infrastructure in order to initiate rehabilitation and maintenance of water supply systems in response to reported data.

It is important to note that although the programme objectives were agreed, the detailed terms of reference were modified as a result of infrastructural changes in the Ministry of Health which continued throughout the 3-year period 1985–88.

The Ministry of Health first underwent a radical reorganization aimed at decentralization of administration to hospital areas. In 1986 the 16 health regions in Peru ceased to represent groups of hospital areas and were replaced by 56 health areas. The health areas corresponded closely to the previous hospital areas. These changes contributed to delays in implementing the pilot programme and were in any case partly reversed in 1987 when the independent areas partly reverted to regions and became department health units, again answerable to Huancayo in the case of the Department of Junin.

Additional changes affecting the programme involve a redefinition of the role of DIGEMA and the separation of the Division of Rural Sanitation (DISAR) from the direct supervision of DIGEMA. This involved the transfer of a majority of DISAR engineering staff from Lima to the health areas and the loss of its central laboratory facilities. More importantly DIGEMA became the Technical Division for Environmental Health (DITESA), acquired the embryo of a central reference laboratory and could be more clearly identified as a separate national surveillance agency since DISAR was now an independent agency called DISABAR, responsible for the construction of water systems.

Until 31 December 1985 Peru was divided into health regions administered from regional capitals. Huancayo was responsible for six hospital areas which made up the XIIIth health region (see Fig. 2.3). It was therefore logical that the Huancayo Hospital and Health Office should provide the main laboratory and administrative base for the local project coordinator.

The Department of Junin was selected for this pilot programme because the Peruvian national plan for rural water supply (PNAPR) was initiated here in 1960–61 and water supply coverage was more advanced than in many other departments.

The XIIIth health region was 70,487 km^2 in area and included all of the Department of Junin and parts of Huancavelica and Pasco. These departments occupy a central position in Peru in the Andean highlands; bounded on the north by the Department of Huanuco, to the south by Ayacucho, to the south-east by Cuzco, to the north-east by Loreto and to the west by Lima and Ica (see Fig. 2.3).

The pilot project study area included 12 provinces in Junin comprising zones of high sierra and high jungle. By far the highest proportion of the population (50 per cent) is located in the fertile valley of the Mantaro river in the high sierra

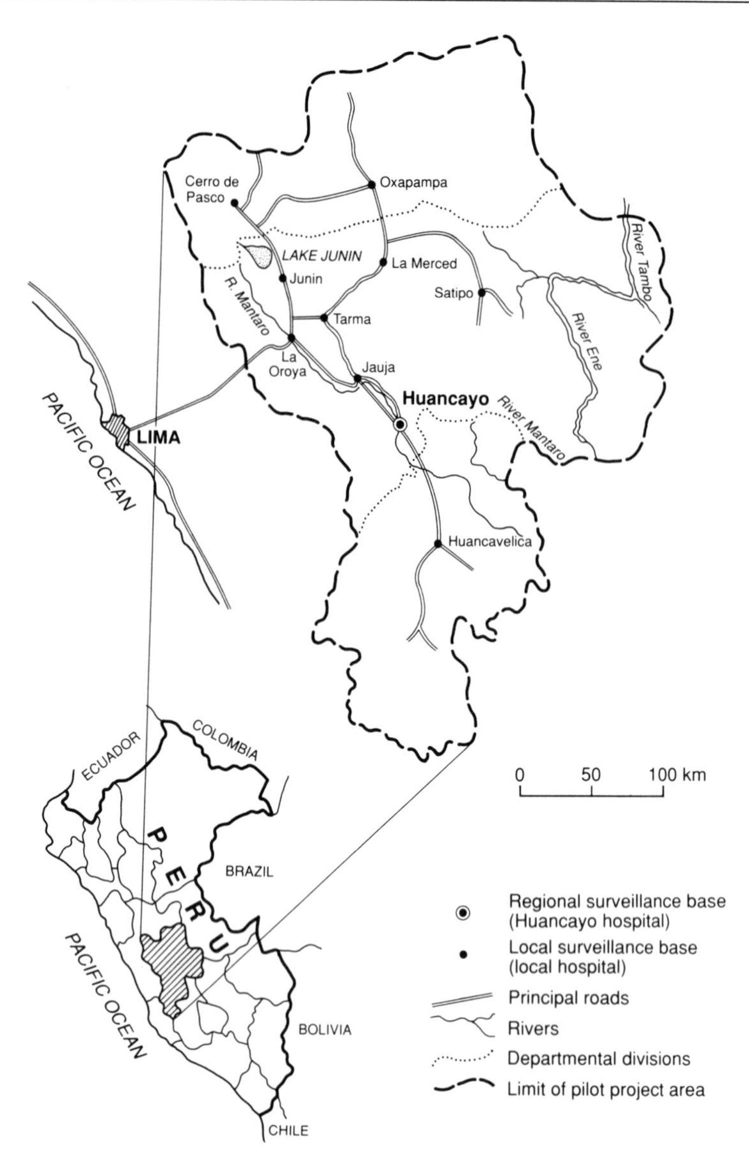

FIG. 2.3 Central region (XIIIth Health Region) pilot project in Peru.

at about 3000–4000 m above sea-level. The Mantaro river together with the rivers Ene, Perene and Palcazu form important headwaters for the Amazon. The Mantaro is already included in a global water quality monitoring network (GEMS/WATER) operated by WHO and UNEP. This was readily justifiable in view of the grossly contaminated nature of the Mantaro which is polluted by a complex of mines and refineries in the Oroya and Morococha areas as well as by sewage pollution from the town of Oroya itself. Fortunately, spring water is used intensively for drinking water purposes by the small communities, but some actually use the Mantaro river or one of its tributaries. The larger communities, not included within this study, are more dependent on surface-water sources.

2.4 The Zambian project

In May 1984, a WHO consultant visited Zambia in order to select a suitable location for the development of a project on control of drinking water quality in a typical rural area of Africa. The project was to be supported with funds from a number of sources including WHO, UNEP, the development agencies of Norway (NORAD) and Denmark (DANIDA) as well as the Government of Zambia.

In due course the Zambian project on the control of drinking water quality was located in the Mongu district of Western province.

In 1984, the Permanent Secretary of Western province established the Provincial Water, Sanitation and Health Education Committee (the WASHE committee). Directives were given to the district executive secretaries in the six districts of the province to set up district WASHE committees and to health assistants to form similar committees at their local level. It was envisaged that the proposed water quality control project would be relatively easily integrated within the WASHE programme. The provincial and district authorities therefore welcomed the project. Since 1980 NORAD had provided support for the construction of shallow wells with windlass, and drilled wells. Throughout Zambia's Western province the water quality characteristics in most wells and boreholes in the province were not known to the Ministry of Health. Therefore a programme of sanitary inspections, water sampling and analysis was considered generally desirable. NORAD had a modest laboratory for water analysis which could be converted into an effective water quality monitoring and surveillance facility (Utkilen and Sutton 1989).

Zambia's Western province is divided into six districts: Mongu, Sesheke, Senanga, Lukulu, Kaoma and Kalabo. The province constitutes a vast plateau of average elevation 1000 m, declining in altitude to the south-east. Most of the province is a vast sandy plain intersected by the flood plains of the Zambezi and its tributaries (Fig. 2.4).

The road network is undeveloped except for the tarred national roads linking Mongu to Lusaka and Sasheke with Livingstone and Ongu. The feeder roads are mostly sandy, uneven, not maintained and suitable only for four-wheel-drive vehicles.

The district has a population of about 120,000 of which approximately 80,000 are rural. The majority of the water sources are traditional water-holes, while the remainder include streams, canals, protected wells and boreholes. Surveys undertaken by NORAD have demonstrated that approximately 45 per cent use protected sources and 10 per cent use surface water (dry season figures). The wet season lasts from October to May, and the annual rainfall is approximately 770 mm per annum. The average air temperature is approximately 30 °C.

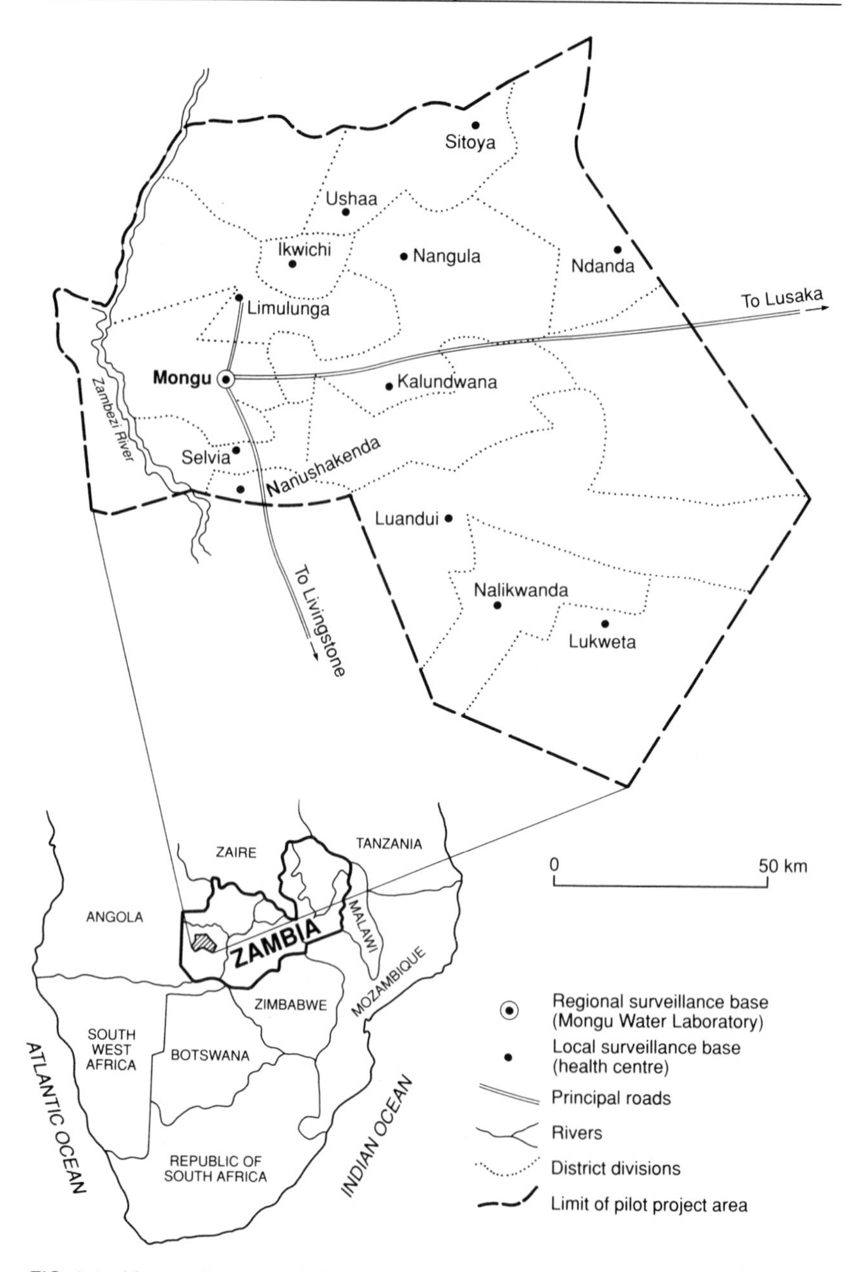

FIG. 2.4 Mongu pilot project in Zambia.

CHAPTER 3

Infrastructure of the Pilot Areas

3.1 Institutional components

The WHO Guidelines and the WHO Surveillance Monograph clearly recommend that: 'The water supplier and surveillance agency should be separate bodies and independently controlled' (WHO 1976a, 1984/85).

This recommendation rarely produces problems in large urban conurbations where quality control is common. It is far more difficult to develop a cost-effective institutional framework for surveillance, maintenance and quality control of rural water supplies where the great majority of these are not self-financing and not operated as commercial activities.

Late in the IDWSSD it was admitted by the major international development banks that construction of water supply systems during the 1980s had progressed without a corresponding development in operation and maintenance capacity, particularly for rural water supply. Official government policy in many countries is that these small supplies should be operated and maintained by the community. Consequently, operation, maintenance and quality control is at best disorganized and most often absent. It is clear that lack of maintenance leads to increasing health risks. It is therefore unfortunate that many developing countries place the burden of both rural water supply construction and surveillance on their ministries of health. We shall therefore examine how the countries of the pilot projects are endeavouring to develop their institutional capacity to control the health risks attributable to drinking water.

3.1.1 Indonesia

Although Indonesian legislation established in 1975 and subsequently in 1977 (Government of Indonesia 1975, 1977) on water resources and waste water, clearly indicates that the Ministry of Health is responsible for surveillance of water related to health, there was not at that stage, nor up to 1988, a defined strategy for implementation of water surveillance at the national level. None the less the Directorate-General for CDC and EH, which was established in 1985, is subdivided into five directorates of which that for environmental health is primarily concerned with sanitation, while the directorates each concerned with water surveillance (surface water, groundwater and waste water) are separate (Fig. 3.1).

A key infrastructural issue is the WHO Guidelines recommendation that: 'monitoring for the routine control of drinking water is the activity of the water supplier; separate checking should be carried out by the surveillance agency; it is highly desirable that the two agencies be separate bodies and independently controlled'.

In the first place, in Indonesia, there are not two independent agencies – one for water quality control and one for surveillance of rural drinking water – and it is most unlikely that government can commit the resources in the medium to long term for routine quality control of millions of small rural installations. This is the reality for a great number of developing countries where the Ministry of Health has the responsibility for rural construction and surveillance. What is important during the initial institutional development is that the task of health risk assessment is taken seriously by whichever agency is responsible for surveillance and/or quality control.

Secondly, the infrastructure for periodic surveillance uses, in the case of rural Indonesia, the same environmental health infrastructure and hence the same staff committed to the installation of rural water facilities, i.e. the Ministry of Health sanitary technicians (Figure 3.1).

Thirdly, for the many millions of family and neighbourhood water sources, there is no local water supply administration. This means that the development of corrective and improvement strategies will present a most intractable problem requiring massive public awareness programmes.

Prior to the initiation of the Gunung Kidul pilot project the provincial and district health service offices were playing a passive role in surveillance. The provincial laboratories often received samples collected exclusively by the urban water authority or occasionally by private individuals. Staff from environmental health sections were rarely involved. To develop an active surveillance programme for the pilot project area it was necessary for the Yogyakarta provincial environmental health section to plan a phased programme of sanitary inspection and sampling by their own sanitarians.

3.1.2 Peru

Prior to 1986, the Ministry of Health divisions responsible for the construction and surveillance of rural water supplies were located within the Directorate for the Environment. However, a new organizational framework was established in the Ministry of Health in early 1986. This effectively separated the surveillance agency, i.e. the Technical Directorate of Environmental Health (DITESA) from the Directorate of Basic Rural Sanitation (DISABAR) which is responsible for the construction of rural water supply systems. This was an important strategic change since the Ministry of Health then conformed more closely to the WHO recommendations that 'the water supplier and surveillance agency should be separate bodies and independently controlled'. The change would be complete if responsibility for construction and administration was separated from the Ministry of Health completely, for instance if DISABAR were amalga-

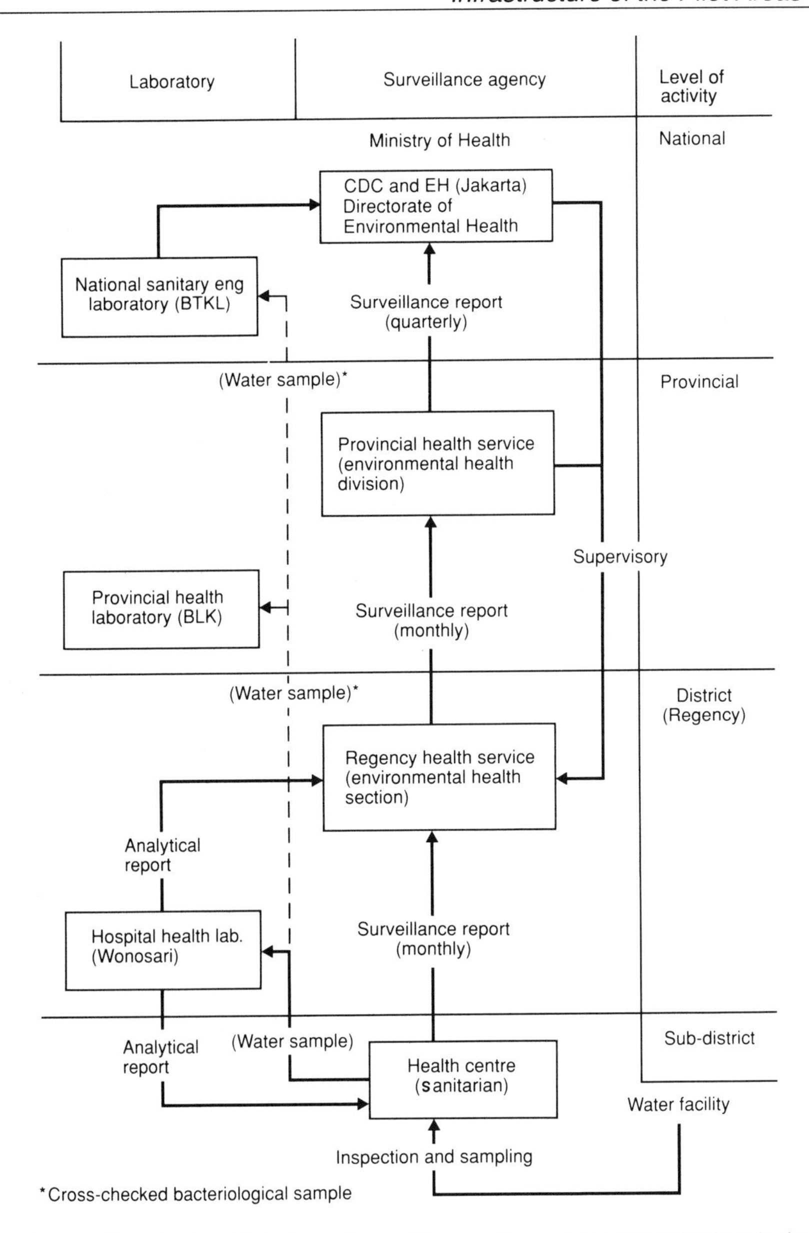

FIG. 3.1 Organizational links for water quality surveillance and communications in the Gunung Kidul project (Java). Note the low level of linkage with water authority at this stage.

mated with the National Water Authority (SENAPA). However, the important point for DITESA is that as the surveillance agency its roles in drinking water surveillance could be defined as follows:

(a) to advise, at the highest level, on policy and strategy which will ensure the development of sustained supplies of safe drinking water;

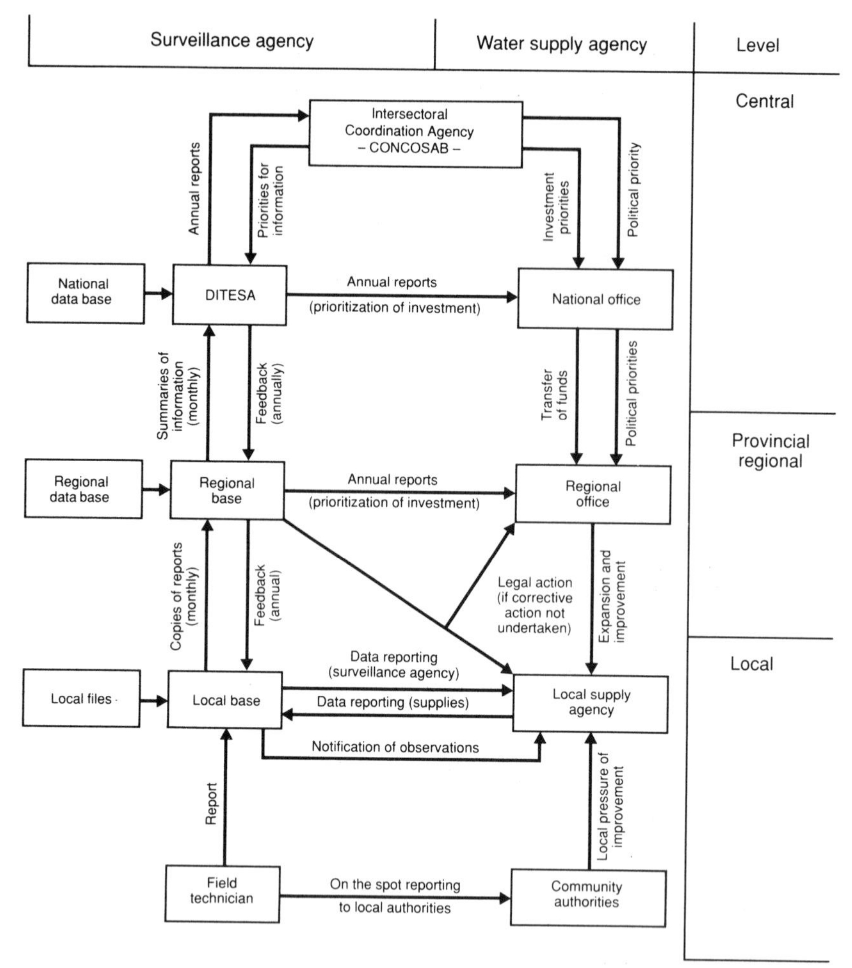

FIG. 3.2 Organizational links for water quality surveillance and communications in the Peruvian project. DITESA = National Surveillance Authority + Central Reference Laboratory; regional base = Directorate of Environmental Health + regional laboratory; local base = surveillance team + water test kit.

(b) to formulate and revise technical standards for the control of drinking water quality;
(c) to supervise, control and evaluate the work of operator level quality control staff and of surveillance staff;
(d) to promote the development of water quality control at the health area level;
(e) to promote and advise on the implementation of water quality control and water surveillance laboratories;
(f) to support and supervise the training of quality control and surveillance staff.

These roles were also stated objectives in the terms of reference of the pilot programme and it was therefore appropriate that a senior engineer from DITESA should coordinate the programme in the region.

Another important change which resulted from reorganization is that DITESA assumed direct responsibility for the laboratory conducting water and soil analysis. This retained intact the concept in the original project proposal for the development of a national central reference laboratory for water surveillance controlled by DITESA.

A policy of decentralization to the health areas in January 1986 removed the regional level of administration and left in considerable doubt the feasibility of implementing and staffing a regional reference laboratory. This situation remained under review during 1986 and thus only the bottom tier of health area laboratories was equipped for basic level surveillance at that time (see Fig. 3.2). However, following a degree of recentralization in 1987, it was decided to establish a regional DITESA laboratory with responsibility for analysis in support of water surveillance in Huancayo.

In 1988 the administrative arrangements of the Ministry of Health were still undergoing change, and the regional reference laboratory was functioning at the initial level.

3.1.3 Zambia

The commitment of the Government of Zambia to water quality control and surveillance may be assessed with reference to a long-standing proposal: 'Water quality monitoring scheme in Zambia' originally prepared by the National Council for Scientific Research (NCSR) in 1982. This proposal envisaged the establishment of a network of water quality laboratories with trained personnel based in existing national and local facilities including those of the NCSR, the Ministry of Health and the municipalities. The network would provide the basis for a surveillance capacity in Zambia because it recognized a formal role for the Ministry of Health. The laboratory base of the water quality surveillance programme in Mongu was situated within the premises of the Department of Water Affairs (DWA) in Mongu township. Both the location of the laboratory and its equipment were largely on the recommendation of a consultant provided by the WHO. The original conception of the laboratory was that having been fairly inactive as a DWA facility, it would be updated using both UNEP/WHO and NORAD funds, staffed by the Ministry of Health and managed within the framework of interministry collaboration in the water sector in Western province. The forum for this collaboration was the WASHE programme, which had both provincial (P-WASHE) and district level coordination (D-WASHE). It was logical to assume before the commencement of the project that the surveillance activities of the laboratory would be managed within the framework of the Mongu D-WASHE (see Fig. 3.3).

The establishment of a 'surveillance' laboratory within the premises of the main water supply agency in the province could be considered to represent something of a departure from the principle of independence described in volume 3 of the WHO Guidelines on Drinking Water Quality (WHO 1984/85) which recommends a clear distinction between the roles of the water supplier and the surveillance agency: in this case the DWA and the Ministry of Health respectively. However, considering the alternative (the pathology laboratory of the Lewanika Hospital in Mongu) and recognizing the preproject understanding that the Ministry of Health should retain a direct interest in the management of the laboratory through its participation in Mongu D-WASHE, the reasoning behind the eventual choice of laboratory location was understandable.

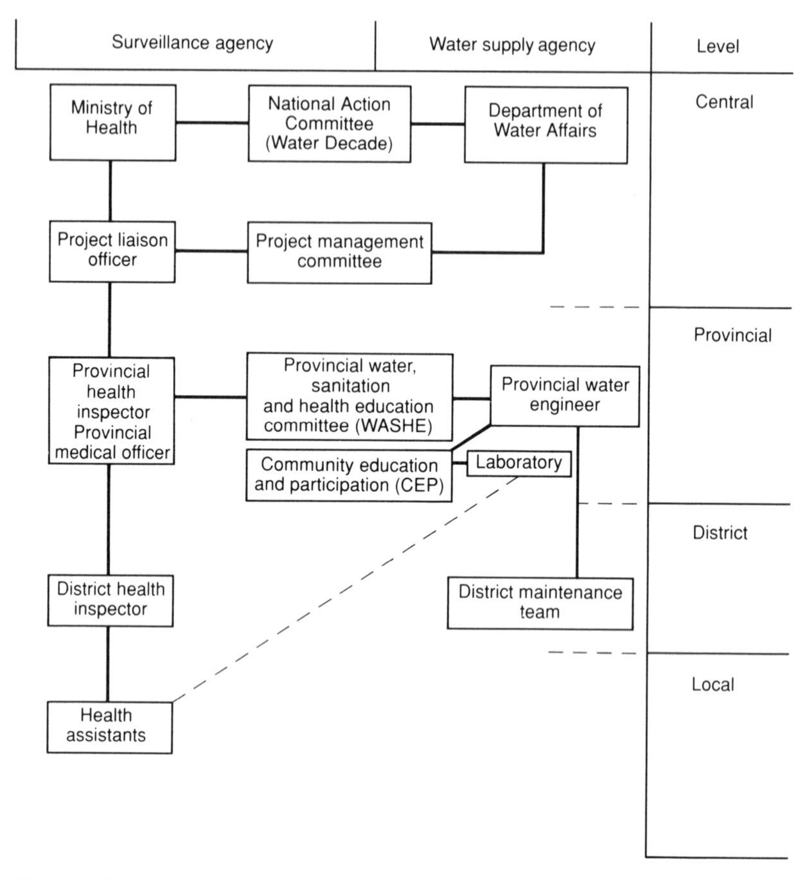

FIG. 3.3 Organizational links for water quality surveillance and communications in Mongu, Zambia. *Note:* the district-level water, sanitation and health committee did not function and is therefore not shown in the scheme.

3.1.4 Comparative evaluation

In both Indonesia and Peru the Ministry of Health has responsibility for water supply construction for small communities, whereas in Zambia this is the responsiblity of the DWA. When we compare the trained capacity for construction, important differences emerge. Thus for a total population of 20 million (40 per cent rural) Peru has over 50 graduate engineers supervising rural construction. By contrast Indonesia has 170 million people (>70 per cent rural) and only 5 or 6 engineers working in point source rural water supply. The policy and emphasis of construction are correspondingly different; in Indonesia and Zambia the focus is on point source systems without distribution networks, whereas in Peru the great majority of supplies are piped.

Consequently, engineering support for the programme of water quality surveillance in the Peruvian project was relatively strong and comprised three principal elements (WHO 1989a):

1. Routine construction and rehabilitation undertaken within the framework of the PNAPR which is the responsibility of DISABAR.
2. Demonstration projects in water treatment plant construction and rehabilitation including gravel prefiltration, slow sand filtration and disinfection.
3. Routine repair of systems by sanitary technicians trained by the programme.

In each of the three countries the WHO Guidelines and the IDWSSD provided an impetus for developing rural water quality control and surveillance. This was effectively absent prior to 1980 although legislation was already in place. It may be concluded from the above that rural water surveillance and quality control present major administrative problems. The priority up to the commencement of the pilot projects had been to attend exclusively to the problem of urban piped supplies which serve a minority of the population. The rural surveillance problem had to be given serious consideration if the investments in water supply for health improvement were to be demonstrated and sustained. It was clear, however, that rural surveillance could not be treated in isolation. It has been proposed that a national surveillance infrastructure be developed in each country, and it is hoped that the pilot projects will demonstrate how to include the rural supplies within a cost-effective system of sanitary inspection, analysis and control.

Surveillance is extremely rare in Indonesia. Within the entire environmental health sector there are only eight engineers and not all of these are dedicated to rural water supply which services well over 100 million Indonesians. The engineering aspects of rural water supply are therefore almost entirely undertaken by health controllers supported by sanitarians trained in appropriate technology but not in conventional sanitary engineering. Personnel with Masters' and Bachelors' degrees in public and environmental health also play a significant role in the environmental health sector.

The pilot Indonesian surveillance project was placed under the general management of the provincial chief of environmental health based in Yogyakarta, who is a health controller. The chief of environmental health is responsible for supervising the installation of unpiped rural water. In Peru this same post was filled by a sanitary engineer who was also responsible for rural water supply construction.

3.2 Organizational linkages

It can be seen therefore that, although in each of the three countries the pilot projects were new initiatives, they depended upon the existing infrastructure albeit with some modifications. Since it was hoped that these initiatives would serve as a model for national and international replication it was therefore also pertinent to examine the infrastructure at national, regional and local level. This can best be summarized by means of organigrams (Figs 3.1–3.3) which demonstrate the communications between and within each level. These figures may be compared with Fig. 1.3.

3.2.1 Indonesia

In the case of Indonesia, the hierarchical infrastructure of the environmental health sections of the Ministry of Health at national, provincial, regency (district), sub-district and village level is shown in Fig. 3.1, which also shows the channels of communication used by the pilot project. It is well suited to the development of a successful water surveillance agency and the communication system for reporting data is described in a document titled 'Control system of drinking water quality and water pollution'. The National Reference Laboratory, however, is at present outside the jurisdiction of the Director General of the CDC and EH and needs to be relocated within it to provide for an integrated system of communication. In addition there are two other sanitary engineering laboratories that act as regional reference laboratories in Java, at Surabaya and Jakarta.

The Indonesian provincial level laboratory infrastructure is already in place to serve the urban populations in most provinces. The provincial laboratories (Balai Laboratorium Kesehatan, BLK) may be physically associated with the provincial hospital or occupy separate premises. Their main function is clinical analysis but they normally include separate labs for analysis of environmental samples including water. This is the case in Yogyakarta provincial laboratory which is classified as a type A laboratory indicating that it carries out the widest range of activities. The statutory reporting requirement is for 33 water parameters to be analysed. This includes 5 physical, 26 chemical and 2 bacteriological tests.

At provincial level the pilot project was managed by the provincial chief of environmental health who is also the general manager for the installation of unpiped rural water supplies; local day-to-day project coordination at the district level was the responsibility of a university science graduate in environmental health. She liaised with the district team of 21 sanitarians and 4 laboratory staff in the district hospital.

What was lacking at the start of each of the pilot projects was a model and *modus operandi* for water surveillance at the rural district level. Management of water quality surveillance at the regency (district) level in Indonesia is the responsibility of the Department of Health Office and the executive agency is its environmental health section. At the district level, which includes district capital towns of tens of thousands of people, the infrastructure for surveillance had not been established and it is rare for environmental health staff to inspect water supply facilities or refer samples to the provincial laboratories. It was therefore proposed that the district level hospitals both in Indonesia and Peru, which have basic clinical diagnostic laboratory facilities, should set aside part of the laboratory area for essential water analysis. In the case of Indonesia the district capital hospital of Wonasari was equipped for this purpose. In Peru, each of the nine hospitals in the health region of Huancayo was equipped with field portable DelAgua water test kits, whereas in Zambia the water quality monitoring facility was limited to the DWA laboratory in Mongu.

At sub-district (*kecamatan*) level in Indonesia there is usually at least one health centre. Thus there are 21 health centres for the 13 sub-districts of the project area. The district capital provides the administrative centre for environmental health sanitary staff. Most sanitarians are based in a health centre, so that each health centre has a sanitarian but they never have laboratory facilities. For the pilot project the more remote health centres were equipped with DelAgua portable water test kits for essential analysis which was carried out by trained sanitarians.

Small community, neighbourhood and family water sources make up the great majority of water facilities in Indonesia and these are the responsibility of the Ministry of Health both for construction and surveillance. At rural level the national water supply policy to promote this type of small unpiped supply poses fundamental problems for surveillance. These problems are common to many other countries where the health authorities are shouldering the burden of rural water supply, but they are particularly difficult in Indonesia both because of the continuing rapid population growth and the many millions of point source facilities which already exist.

3.2.2 Peru

The Peruvian pilot project was a joint initiative of the Directors of the National Offices of DISABAR and DITESA. As indicated in Fig. 3.2, their respective offices liaised directly, as well as through the official National Coordinating Council for Environmental Health (CONCOSAB). This ensured that, from the beginning of the project, there were mechanisms available to develop remedial action strategies using the information generated by the surveillance activity. The existing infrastructure also ensured that beyond the pilot project there were in-built mechanisms to use the data for the strategic planning of investment for the preservation and improvement of water supply. This planning would become necessary when the project developed into a regional and eventually a national programme.

The concept of a national surveillance team for implementation and management of surveillance was developed within the project and had the advantage of bringing together coordinating and supervisory staff within DITESA from a variety of appropriate disciplines. By the end of the pilot project the team included a national director/manager, sanitary engineer, chemist, biologist and laboratory technologist. There was the additional advantage that DITESA already housed the Central Reference Laboratory and subsequently the surveillance data base.

The supervisory and technical support provided by DITESA was progressively strengthened during the pilot project so that the technical corporation consultants of the UK ODA could withdraw from field supervision once a routine pattern of reporting and supervision had been established.

At regional level in Huancayo liaison between DITESA and the regional health director and the director of environmental health existed before the project began. The regional director for environmental health is part of the DITESA infrastructure and is also the sanitary engineer responsible for rural water supply construction. He is therefore the chief of the sanitary technicians some of whom were delegated to surveillance activities. The regional surveillance base staff, data records and laboratory were located in the Huancayo hospital complex in a building which housed the Rural Sanitation Division.

Within the period of the pilot project the regional surveillance laboratory was developed to the initial level (a) indicated in Fig. 4.4. This provided laboratory facilities for routine urban and rural surveillance and was used as the centre for distribution of testing consumables and supplies to the local level hospitals in the health areas.

The surveillance team concept was used again at hospital health area level and developed to allow two to six technicians to work together at least on the surveillance diagnostic. The team included mixed skills and was formed from those staff based at the hospital. It included at least a coordinator, who was sometimes a hospital laboratory technician and sometimes a sanitary technician, and a driver. The tasks for each member of the basic surveillance team are those listed in Table 3.1. In addition to the tasks listed in the table the most effective teams carried out remedial action on the spot and in one team this was done by the driver who was also a plumber. More often though the remedial action was done by the sanitary technician and the teams were equipped with tool kits for this purpose.

On completion of a water supply sanitary survey, which is always conducted with a member of the community's administrative committee, the field sanitary technician provided on-the-spot summary reports to the community authorities. The report indicated what remedial action was needed and, as mentioned above, the survey team could sometimes carry out repairs. The full sanitary survey reports were filed in the local surveillance base at the hospital and routinely copied to the area water supply engineer and the regional director of environmental health. At the end of the financial year the data were summarized and analysed to provide a priority list of the systems requiring remedial action as described in section 7.3.8. Consolidated report forms were also transferred periodically to DITESA in Lima.

It can be seen from the above and from Fig. 3.2 that effective top-down

Table 3.1 Summary of tasks ascribed to each member of the Peruvian surveillance team

Tasks	Coordinator	Sanitary technician	Laboratory technician
1. Financial planning	+		
2. Chronogram of visits and administration of vehicle use	+		
3. Inventory of equipment and consumables	+		+
4. Prepare and dispatch monthly résumé of results	+		
5. Manage data base/archive	+		
6. Prepare raw data sheets	+/−	+	+
7. Interview and communicate with administrative committee		+	+
8. Execute sanitary inspection		+	+/−
9. Execute water analysis		+	+
10. Maintain laboratory equipment and prepare reagents			+

supervision was developed by DITESAs consultants, although this required several years to establish effective transfer of responsibility to the DITESA team. Equally a bottom-up and cross-agency reporting system was developed during the same period.

3.2.3 Zambia

In the case of Zambia, the organizational structure of the water quality surveillance project must be considered in the content of existing interministry collaboration. The national management and liaison remained relatively consistent, the local managerial staffing did not. The NORAD support for the water supply sector in Western province was an important influence in the support received by the project at local level. The organizational structure for surveillance activities is illustrated in Fig. 3.3.

The key lines of communication on matters relating to the project were established at both national and local level, but communications between the two were poor. However, with the appointment of a new provincial health inspector in 1988 improvements in communications occurred within the local Ministry of Health structure.

As a result of the poor communication between different levels in the early stages of the project there was no direct technical exchange between the Mongu laboratory and Lusaka. The chief analytical chemist of the Lusaka laboratory established this contact in late 1987. There were no suitable laboratory facilities within the Ministry of Health at provincial or district levels to which the water quality surveillance could relate.

Engineering support for water quality improvements in Mongu district was provided primarily through the rural water supply programme of the DWA, supported by NORAD. It should be noted that DWA and NORAD staff did not consider that the results of the surveillance programme contributed directly to the assessment and prioritization of water supply schemes at district or provincial levels. However, health assistants employed by the Ministry of Health, together with local political leaders (ward chairmen), were important in assisting and influencing the activities of the WASHE Community Education and Participation (CEP) programme at health catchment area level.

The CEP made recommendations for new engineering interventions by the use of a selection criteria form for assessing village needs. Typically a score of 60 (out of 100) was considered the minimum for inclusion in the rural water supply programme of the DWA. The need for remedial and rehabilitation interventions was assessed in the course of annual inspections undertaken by DWA staff in consultation with health assistants.

3.3 Project resources and facilities

3.3.1 Human resources

The human resources for each of the pilot projects may be identified at coordinating, supervisory and executive levels. In Indonesia, national supervisory support to the pilot project was provided by means of periodic visits from staff of the surveillance section of the Subdirectorate of Surface Water of CDC and EH as indicated in Fig. 3.1. Graduate engineering support for water supply is really only available for piped supply construction to larger communities. The engineering aspects of the majority of rural water supply, which is unpiped, are almost entirely undertaken by the health controllers supported by sanitarians trained in appropriate technology but not in conventional engineering. Personnel with Masters' and Bachelors' degrees in public and environmental health also play a significant part in the environmental health sector. The pilot surveillance project was placed under the general management of the provincial chief of environmental health based in Yogyakarta, who is a health controller. The chief of environmental health is responsible for supervising the installation of unpiped rural water facilities and is normally a qualified sanitarian who has undertaken 2 years of further studies to become a health controller. Similarly in Peru, the regional director for environmental health was the nominated director for the project. He too had the dual responsibility of rural water supply construction and surveillance; and the undesirability of this has already been mentioned. In Zambia, by contrast, the titular director of the project could be said to be the provincial medical officer. It was only in Peru, however, that a project director was appointed at national level, in DITESA. He was a sanitary engineer, as was the regional director.

In Indonesia, local day-to-day project coordination at the district level was the responsibility of a university science graduate in environmental health. She was responsible for water supplies and sanitation and was based in the district health office at Wonasari. The comparable level of coordination in Peru was achieved for each health area by nominating a surveillance team coordinator. In Zambia this could be said to be the district health inspector.

The work-force responsible for execution of sanitary inspection and water sampling was based on sanitarians in Indonesia and Peru and health assistants in Zambia. Health assistants and sanitarians' duties include public health and hygiene education, supervision of solid waste management, sanitary inspection and certification of public places and private restaurants, licensing street vendors, sanitary inspection and sampling of foodstuffs including slaughter houses; zoonoses, pest and vector control, and last but not least, waste water and public water supply facility construction and surveillance. Without demarcation of these multifarious functions it is likely that none of them will be done very professionally. It is obvious that a degree of specialization is essential as well as specific vocational training, which was provided for in the terms of reference of each of the pilot projects.

In Indonesia the district surveillance team comprised 21 sanitarians, each based in a health centre responsible for sanitary inspection and water sampling part-time. Four laboratory staff – 3 based in the Wonosari hospital and one in the regency health office – were responsible for water testing. In Peru a surveillance team was based in each of the 9 principal hospitals spread through the 6 health areas. In all 35 staff, mainly sanitarians, were trained and about 30 were active throughout the project in both rural and urban surveillance. In Zambia, 18 health assistants, mainly based in health centres, were trained for the project in the 14 health catchment areas which sent representatives to the preliminary training course; 8 took samples regularly in 1987, 11 undertook sanitary inspections and 3 were inactive. In calculating numbers to be trained a 10–20 per cent allowance should be made for wastage over 3 years.

It can be seen that the staff in Indonesia and Zambia were decentralized to the health centre level with respect to the surveillance function, whereas in Peru the sanitarians functioned as teams coordinated at the area hospital level (Table 3.1).

3.3.2 Physical resources

The physical resources necessary for operating the pilot project may be taken to include offices, laboratories, health centres, transportation and equipment. In Indonesia and Peru the principal physical bases for the projects were the existing hospital laboratories and health centre offices. In Zambia, the base was the Mongu district water laboratory. None of the projects had new buildings. Each used existing space but the district laboratories all required service improvements and equipment as described in the basic laboratories section of the WHO manual on water laboratories (WHO 1986). This was also recommended by an earlier WHO guideline (WHO 1976a).

Thus all projects complied with the recommendation, but in the case of Peru the project simultaneously established a basic regional laboratory and local bases for field testing by all nine survey teams as shown in Fig. 3.2. Indonesia likewise adopted this structure but with fewer health centres equipped with field test kits. By contrast Zambia had to transfer all samples to the one district laboratory and its sampling programme was thus limited by transport problems.

The availability and/or cost of transport posed major constraints on all three pilot projects. In Gunung Kidul most of the inspection and sampling was done by motor cycle. In Peru nine Ministry of Health Land Rovers were rehabilitated and maintained under the terms of the bilateral aid agreement; their maintenance and fuel consumption were major components of operating costs and because they were properly maintained were shared with other health projects. Furthermore the Peruvians had the greatest distances to travel over difficult terrain. In addition, three motor cycles were donated to the Peruvian pilot project by WHO/UNEP. Those were used effectively in the jungle areas but not on the highland mountain roads. In Zambia no transport was available for coordination nor to most of the health assistants. They delivered samples to Mongu, largely at their own expense, while NORAD provided monthly transport of samples from the most distant health centres. Only two health assistants had motor cycles for transport of samples. The Ministry of Health provided fuel for these but the majority of the transportation budget was not spent.

3.3.3 Financial resources

The financial resources of the projects for the first year are summarized in Table 3.2. Although the investment per head of population in each project is modest (12–77 US cents), it is worth while examining the breakdown of costs line by line, thus:

1. The very low level of the Zambian management and supervisory costs ($US 1350) reflects amounts claimed for supervisory visits by Lusaka-based staff and does not take an account of salaries since these data were not available. However, since the communication between central administration and provincial bodies was poor, and since local management was not established until late in project implementation, the amount is probably a fair reflection of value of input. The comparatively high figure for Peruvian supervision ($US31,600) is equally a fair representation of the involvement of local, regional and national coordinators.
2. The relatively high salary figure ($US11,900) of Peruvian sanitarians reflects the number of sanitarians active in the programme. They were working both in the rural and urban sectors. Since the Peruvians worked as teams which include the sanitarian, lab staff and coordinator, there is no additional cost for lab staff.

Table 3.2 First year budget estimates in US dollars for each of the pilot projects

	Indonesia	Peru	Zambia
National contributions			
1. Salary: management and supervision	12,500	31,600	1,350
2. Salary: Sanitarians	5,000	11,900	3,820
3. Salary: Lab staff	2,000	—	?
4. Training workshop	3,000	500	2,800
5. Operating costs	5,000	22,000	2,800
6. Evaluation workshop	—	—	2,875*
	28,000	66,000	13,645
Donor contributions			
7. Capital: Equipment and transport	4,900	38,000	6,000
8. Operating costs	5,000	12,000	—
9. Launch and training workshops	25,000	4,000	45,300
10. Consultancies	20,500	50,000	27,800
	55,400	104,000	79,100
Total available budget ($US)	83,400	170,000	92,745
Total target population	690,000	1,352,000	120,000
Investment per capita (US cents)	12	12	77

* Not spent

3. This means that if lab staff costs are added, then the cost for Indonesia is $US7000. Unfortunately the cost was not available for Zambia. The cost per annum of sanitarian/lab staff can then be compared for each country, thus:

   Indonesia 7,000 ÷ 24 = $US292 per person
   Peru 11,300 ÷ 30 = $US396 per person
   Zambia 3,820 ÷ 8 = $US477 per person

   Although these values might not actually take account of time dedicated to surveillance, they do suggest that teamwork may be most cost-effective since the Peruvian costs incorporate ***urban surveillance as well.***
4. Training workshop costs are all comparable since the costs were Indonesia ($US3000), Peru ($US500 + $US4000 for item 9 borne by the donor agency) and Zambia ($US2800). There is an economy of scale for the numbers trained and an optimal number of participants is 25–30.
5. The high operating costs incurred by Peru is due to costing in of drivers, secretarial and local level administrators.
6. Evaluation workshops were run in the second and third years of the projects

in Indonesia and Peru, but not at all in Zambia. Although it was budgeted for in the first year, the amount ($US2875) was not spent and reflects severe institutional weaknesses.

7. The basic laboratory equipment supplied to Indonesia and Zambia was similar and, indeed, the great majority of equipment and consumables for water testing was donated by UNEP/WHO, NORAD and ODA for each of the projects. The far higher sum for Peru ($US38,000) was split between field testing consumables for 3 years and for 9 teams ($US12,000), 3 years' Land Rover parts ($US10,000) and motor cycles ($US4000).
8. Operating costs underwritten by the donor agencies for Indonesia and Peru included travelling expenses for national and local supervisory staff and report printing. In the case of Peru, however, the main element was for fuel, lubricants and repairs of vehicles and travelling expenses for field staff. These donor items were phased out in the third year of the project and assumed by the national government.
9. The costs of the national launch of the projects is particularly high in the cases of Indonesia and Zambia and disguises the fact that participants attended from other countries in the region and thus includes airfares and international speakers contracted by WHO and DANIDA. By contrast the Peruvian project did not have a national launch and was limited to a training workshop within the pilot area.
10. The consultancies provided the main impetus and technical advisory support for the implementation of each of the projects; but whereas Indonesia and Zambia used consultants for short-term inputs, Peru relied on Del-Agua consultants to develop and co-manage the project in the medium term. This represents the most costly of all inputs but it was also the most innovative since almost all of the conceptual developments in surveillance reported here derive from this source.

3.4 Community involvement

Finally, it is important to consider the role of the community (user), particularly in relation to the operation, maintenance and quality of control. In Indonesia for the many millions of family and neighbourhood water sources, there is no water supply administration. This means that the development of corrective and improvement strategies will present a most intractable problem requiring massive public awareness programmes.

By contrast, the village communities agreeing to accept a water supply system constructed by the PNAPR in Peru are required to establish a proper administrative committee of five comprising president, secretary, accountant, speaker and operator. The field technician/sanitarian provides on-the-spot reports to this community authority (see Fig. 3.2).

In Zambia, the WASHE philosophy was that communities were to be made responsible for their own supplies, and educated in cleaning and more technical maintenance of their own protected sources. This approach was developed for new sources in 1986, but only covered the backlog of the quality control project, which ran from November 1986 to November 1987. The two activities were therefore mutually beneficial, if not as well coordinated as they might have been.

There is now a tradition in many developing countries of community involvement in the construction phase of water supply development. Once they have contributed an intensive effort, it is far more difficult to sustain the interest of the community for operation and maintenance and surveillance. None the less in recognition of professional manpower deficiencies volume 3 of the WHO Guidelines specifically recommends community involvement in surveillance even though this may place heavy demands on the sanitarians who will be required to train the community in these activities. Options proposed in volume 3 to implement community surveillance include:

1. selection of volunteers;
2. payment of a stipend, by the community, to an operator to carry out routine tasks;
3. payment of a professional.

All of these assume a measure of management and the election of a suitable committee.

Indonesia is officially basing its surveillance activities at the lowest level, i.e. point source private supplies, on the first option, with support from sanitarians for the larger community supplies. At this lowest level of the village (cadre) volunteers, they will be trained to carry out simple sanitary inspection only, as decribed in Chapter 5.

It should be noted that the Indonesian system is different in supervision for remedial action at sub-district, i.e. small community level. The other two countries, however, established remedial action within their pilot project strategies; but in contrast their overall monitoring infrastructure was weak, since, unlike Indonesia, they had not established a provincial health laboratory network for water quality monitoring prior to their pilot projects.

It may be concluded that in the early development of national surveillance programmes countries will sometimes have to depend upon institutions which are already involved in water supply construction and not on the recommended ideal of an independent agency. Recognizing the difficulties involved in establishing any form of independent monitoring institution for rural communities it is pragmatic to recommend that cooperative action with the community, by *active* agencies, is more important than independence since the overall objective is improvement of water supply services.

3.5 Training support

3.5.1 Training programme in the pilot areas

The general vocational training for environmental health officers (sanitarians) worldwide varies from a few months to several years. In Indonesia there are 13 colleges for sanitarians providing a one-year full-time general course. In Peru, the training of sanitarians, by the Institute of Public Health, has been reduced in the 1980s from 2 years to less than 6 months. The general training of Zambian health assistants is for 3 years in Lusaka. None the less it was considered essential in all three projects to provide supplementary specific training preparatory to implementation of water surveillance.

Table 3.3 Breakdown of staff trained, topics addressed and time allocated in the pilot projects' principal water surveillance training courses

	Indonesia	Indonesia	Peru	Zambia
Trainees	Sanitarian	Lab tech	Sanitarian	Health
Type			and lab tech	assistants
Number	21	4	35	18
Course duration (days)	5	16	10	10
Topics addressed		*Hours allocated*		
1. WHO volume 3 Guidelines	1.5	1.5	2	4
Surveillance methods	1.5	1.5	1	2
2. Water-related disease	—	0.75	2	2
Faecal indicator theory	—	0.75	1	—
3. Project outline plan	0.75	1.5	1	2*
Review existing data	—	—	1	—
4. Water treatment and supply	—	—	10	7
Operation and maintenance	—	—	5	—
Remedial action and rehabilitation	—	—	2	3
5. Community participation	—	—	2	4
Health and hygiene education	—	—	—	4
6. Water sampling	8	—	2	9
7. Analysis: chemical	1.5	10	3	2
bacteriological	3	21	4	2
reporting results	—	12.5	2	3
8. Equipment, maintenance	—	1.5	3	—
Sterility, hygiene and media	—	11	2	—
9. Sanitary survey (theory)	1.5	—	2	—
Inspection and analysis (field)	—	—†	16	—
10. General data reporting	0.75	—	4	3
11. Project preparatory work	1.5	—	6	3+3*
Inventories, timetables				
Group discussions	3	—	5	—
12. Course evaluation	—	—	—	1
Total contact time (hours)	23	62	76	54

* Separate 1-day preliminary seminar.
† Supplementary 9-day field kit course.

In Table 3.3, the staff trained and the key features of training are summarized and fundamental differences emerge. The most obvious difference was the split training conducted in the Indonesian project. Separate courses were established for the field sanitary staff and Wonosari laboratory staff. By contrast the course in Peru was conducted jointly for the sanitarians with laboratory technicians and they subsequently worked throughout the project as teams (Table 3.1) and carried out simultaneous inspection and analysis to produce a diagnostic of all systems. This approach was partly dictated by the fact that all Peruvian systems studied were piped supplies requiring more time to study in the diagnostic phase than the preponderance of point source unpiped supplies which characterized the Indonesian and Zambian projects. Zambian training was confined to the health assistants since the water laboratory was established with staff specifically trained for water testing.

The duration of the water courses was the same in the case of Peru and Zambia (10 days), whereas the laboratory training alone in Indonesia ran to 16 days. This does not include the additional training for the use of field test kits (9 days). It is clear in retrospect that a 5-day course for sanitarians, as conducted in the Indonesian project was, by comparison, inadequate, and this was also reflected in the limited contact time (23 hours); when the topics covered in each of the courses are compared (see Table 3.3), it can be seen that whole themes were missed out. Thus the Indonesian course provided no element on water treatment and supply, whereas the Peruvian and Zambian courses included 17 and 10 hours respectively. It may have been argued that these elements were less vital in Indonesia because the majority of water sources are untreated and unpiped, but it also excluded operation, maintenance and remedial action which are vital components of overall surveillance strategy. Furthermore, both the Indonesian and Zambian courses had minimal or no time allocated to sanitary surveys! In the case of Indonesia this fact largely wasted the surveillance effort. The disproportionate emphasis placed on water testing in the first 2 years of the project meant that the analytical results could not be related to any particular structural problem which effective sanitary inspection might otherwise have revealed. In fairness it has to be said that all three projects were, at one stage or another, defective in their reporting of sanitary inspection to complement the analytical data. Fortunately, however, in the case of the Peruvian project many of the sanitary inspection problems were sorted out at the preplan stage.

3.5.2 Preparatory training for project staff

The preplan Peruvian project began in 1984 with a 4-day training course, but made the error of focusing primarily on laboratory technicians. Although this short course dealt both with sanitary inspection and analysis it was subsequently obvious that whereas the training in analysis was adequate, the training in inspection was seriously deficient. It was evident from several survey report forms that the technicians had even sometimes failed to identify correctly the component of the system they had sampled. The opportunity was therefore used to revise the WHO model sanitary report forms and, prior to the principal training course, these were field tested in the pilot programme area. These were evaluated by the regional sanitary inspector responsible for preservation together with a consultant. Because the data evaluated in the Peruvian preplan area demonstrated that most systems were poorly maintained and supplied faecally contaminated water, the concept of the primary diagnostic survey was introduced at the training course as the foundation from which to launch routine surveillance. Consequently one of the principal topics addressed in the Peruvian pilot regional training programme was the sanitary survey. This occupied more time than any other component of the course and was equal with all components of analysis (Table 3.3). Neither the Indonesian nor the Zambian projects record any field practice in sanitary inspection and as a consequence these projects, like the Peruvian preplan, were seen as a laboratory exercise in the early stages and failed to provide useful information on which remedial action could be based.

3.5.3 Evaluation and retraining

All three projects suffered from local and national management problems. It was therefore essential to have an interim evaluation of the surveillance projects at which progress could be reported and the technical, as well as managerial, issues could be reviewed and compared. This opportunity was provided in May 1987, by a USSR/UNEP/WHO interregional workshop which dealt with the hygienic training of project field staff, particularly in the light of the sanitary survey deficiencies identified.

In the case of Peru, the local review workshop was held in the pilot regional capital (Huancayo) in June 1987. This provided the opportunity for the survey teams not only to present and evaluate their diagnostic data, but also to plan towards routine surveillance. The occasion was also used to examine formally and certify the surveillance staff who by then had been working for 14 months on the pilot project. Neither the Indonesian nor the Zambian staff were certified although both carried out refresher workshops. The Zambian refresher course was carried out a year after the first training course, whereas the Indonesian retraining did not occur until over 2 years later. This was particularly unfortunate since the sanitary inspection component had to be completely revised and was therefore only applied successfully during the last 6 months of the Indonesian pilot project.

3.5.4 Recommended training activities

The combined experience from developing courses has been useful in assessing what preparatory activities should be undertaken to try to ensure that training is successfully brought to fruition. These preparatory activities are therefore brought together in the training and implementation checklist (Table 3.4). This indicates that training should be considered to be the final stage in planning and the first stage in implementation. It demonstrates bad planning if implementation cannot quickly follow training because without rapid implementation the surveillance staff would soon require retraining. It will be noticed, therefore, in the following checklist, that planning implementation is an integral part of the training course and implementation components occur 17–30 days after the training course.

One of the most important aspects of surveillance work is the coordination of the surveillance team. Belatedly, in the third year of the project, the Indonesian staff organized a coordination workshop, but it is now believed that this should be done most cost-effectively at the first training course. To strengthen teamwork it is invaluable to bring together the coordinators, sanitarians and laboratory technicians to be trained and plan their implementation strategy together. To this end the recommended course is common to the whole team and deals with the integrated aspects of surveillance, including field analysis and planning.

A model for a basic common course for the surveillance team is presented in Table 3.5. It deals first with rural systems beginning with unpiped systems where sanitary inspection may be done without analysis. It is envisaged that this simplest form of investigation will be the core activity for village volunteers (cadres) whom the sanitarians will subsequently train. To inspection is added critical parameter field testing for rural piped systems. This is reinforced for urban piped systems and only essential additional tests are dealt with at this local level of training.

Table 3.4 Surveillance training and implementation checklist

	Actions	Date on which action completed	Countdown (days)
1.	(a) Nominate course coordinator (b) Nominate course secretary		
2.	Agree and clear course dates with district health directors (6 months in advance)		−200
3.	Select sanitary surveillance teams (district coordinators, sanitarians and laboratory technicians) for training from those priority areas where budget permits implementation		−150
4.	Inform trainees about the course dates 4 months in advance		−120
5.	Select and invite local and national trainers and agree on their conditions		−110
6.	Formulate training course syllabus		−100
7.	Identify water facilities for field-work		−100
8.	Select and reserve training centre and accommodation		−100
9.	Draw up training materials list: (a) Prepare summary inventories of local towns' water supply coverage to calculate frequency of water testing (b) Obtain local water supply distribution map to demonstrate fixed sampling plan		−90
10.	Order from national surveillance agency: (a) Sample inspection report forms (b) Ministry of Health Water Regulations (c) WHO Guidelines Vol. 3 and request dispatch 4 weeks before training course		−90

Table 3.4 *cont* Surveillance training and implementation checklist

	Actions	Date on which action completed	Countdown (days)
11.	Formally invite trainees to attend the course and instruct them to bring: (a) Work calendar, to calculate and plan days/month to be worked on water surveillance (b) Local road maps, to plan distances to be travelled (c) Urban water supply works and distribution plans, to plot fixed sample site locations (d) Corresponding population data to calculate frequency and number of samples and inspections to be made yearly		−90
12.	Three days before the course check all equipment and materials at the training centre		−3
13.	Check trainer's requirements for projection and demonstration facilities		−2
14.	Execute training course and agree on chronogram of implementation for each district		0
15.	Each district coordinator prepares action plan with sanitarians and laboratory staff during or immediately after the training course		+7
16.	Equipment and supplies transferred to the district team		+10
17.	Mobilize survey teams and requisition operating costs		+20
18.	Commence inspection and sampling chronogram		+30

Table 3.5 Model training syllabus for basic initial water surveillance course. Joint training for sanitarians, laboratory technicians and coordinators

Day	Topic	Activity	Session
1.	Introduction to course objectives	Introductory	1
	Distribution of course materials		2
	Outline of Ministry of Health Regulations	Theory	3
	WHO Guidelines	Theory	4
	Water-related disease, epidemiological evidence; hygiene and health	Theory	6
2.	Planning surveillance programmes	Theory	1–2
	Sanitary inspection methods	Theory	3–4
	Preplan results, relating risk assessment to bacteriological contamination	Theory	5–6
3.	Detection of faecal coliform; significance	Theory	1–2
	Field test kit; membrane faecal coliform	Demo	3–4
	Bacteriological sampling and hygiene	Demo	5–6
4.	Introduction to Field-work		1
	Sanitary inspection and bacteriological analysis; dug wells and tubewells	Field-work	2–6
5.	Notation of bacteriological faecal coliform results	Practical	1
	Discussion of relationship between inspection and counts	Theory	2–3
	Disinfection theory and technology	Theory	4–6
6.	Chlorine residual and turbidity – test kit	Demo	1
	Reporting appearance, taste and odour	Demo	2
	Field testing pH, Fe, nitrate	Demo	3–4
	Rural piped supplies – sanitary protection	Theory	5–6
7.	Urban water supply – provincial system	Theory	1
	Introduction to urban sampling	Field-work	2
	Inspection and urban sampling for Cl_2	Field-work	3
	Visit distribution public tap	Field-work	4
	Chlorine residual and bacteriological	Field-work	5
	Sampling from 4 points in distribution	Field-work	6
8.	Faecal coliform membrane method at the training centre	Practical	1–3
	Planning sampling programmes, urban	Working group	3–5
	Discussion of preliminary plans. Plenary	Discussion	6
9.	Notation of bacteriological results	Practical	1–2
	Safe disposal of cultures and discussion	Demo discussion	2–3
	Planning of rural surveillance	Working groups	3–6
10.	Presentation of work plans	Working groups	1–3
	Remedial action strategies, rural	Theory	4–5
	Final discussion and course evaluation	Discussion	6

CHAPTER *4*

Surveillance Planning

4.1 Approaches to planning

With hindsight from the three projects it has been possible to develop a model methodology as a series of procedural steps. The WHO surveillance monograph (1976a) provides an overall national framework and describes five levels of surveillance of increasing complexity with primary emphasis on urban systems. It does not deal with institutional problems and detailed planning for implementing surveillance in small communities. Indeed it specifically precludes rural surveillance below level III. By contrast volume 3 of the WHO Guidelines (WHO 1984/85) provides some, albeit brief, planning information for surveillance.

Volume 3 identifies two levels of surveillance for small communities:

1. *Initial:* characterized as irregular surveillance and severely limited.
2. *Advanced:* characterized as having all surveillance control elements fully operational.

No publication provides detailed surveillance project implementation guidelines; therefore it is not surprising that the pilot projects all made serious common mistakes in the early phases. Prerequisites for project implementation are that there should be national government regulations and standards and in the absence of these the aforementioned WHO publications may be used as general guidelines. Another prerequisite is that there should be a lead agency at national level responsible for the institutional development of surveillance strategy.

These institutions have already been described in Chapter 2 and the reasons for the choice of project areas indicated. We therefore take up the planning procedures from this point as a result of the pilot project experience. Three principal components have been identified in the development of a provincial surveillance programme:

1. A *preplan* project which demonstrates the validity of selected methods in a small study area (sub-district).
2. A *diagnostic* study which requires a complete inventory of all community water services and full sanitary surveys of all piped supplies and a

representative selection of unpiped, point source community supplies in every village in the province and pilot project area.

3. *Progression to routine surveillance* requiring the continuing and periodic checking of all piped supplies and inspection of all new and improved unpiped sources.

In the first year of the project it is valuable to work to a checklist of preparatory actions which need to be executed if the programme is to have a reasonable chance of being sustained and successful. The checklist is intended to be a logical sequence although it may not be possible to stick rigidly to the sequence suggested. One useful outcome of the pilot projects was that a project planning checklist was developed, incorporating experience from all three projects. It is presented in Table 4.1.

The Peruvian project was the only one of the three projects to have a clearly defined medium-term strategy prepared in advance of implementation. The Peruvian plan included four phases of implementation, progressive expansion and replication over 6 years as follows:

Phase 1: *Preplan* agreed and begun in 1984 with the support of DITESA in cooperation with DISABAR. Preliminary training of laboratory technicians was to be followed by implementation in selected communities in one department in the central region of Peru. The results of the preplan would be evaluated within 1 year and the lessons learned from this incorporated into a pilot regional programme.

Phase 2: *Pilot regional programme* (1986–87). This included planning to incorporate all the health areas within the health region to execute a surveillance diagnostic. It envisaged establishing a regional coordination centre and basic reference laboratory together with nine surveillance teams each based on a provincial hospital. The surveillance teams were to be trained to implement the diagnostic including full sanitary surveys and water quality analysis. As a result of the diagnostic it was planned to establish a rehabilitation strategy initiated with a series of pilot rehabilitation projects. Phase 2 would end with an evaluation workshop.

Phase 3: *Full central region programme* (1988–89). The phase 2 experience would be replicated throughout the entire central region of Peru to include six departments and provide routine urban and rural surveillance coverage for approximately two-thirds of the population of the country since it included the metropolitan area of Lima. Phase 3 would end with a national evaluation workshop and launch of the phase 4 National Plan.

Phase 4: *Complete national coverage* (1990–95). Phase 3 would be expanded to the northern and southern regions of the country by establishing coordinating centres and regional reference laboratories based on the model established in phase 2.

Three of the four phases are presented schematically in Fig. 4.1. A similar strategy has been prepared for Indonesia as a result of its pilot project and this is shown in Fig. 4.2.

Table 4.1 Pilot project planning checklist

	Period of action	
	Start date	Completion date
1. Preplan project methods selected and tested in small study area by national level professional staff	________	________
2. Preplan project methods evaluated and results reported	________	________
3. Provincial programme preliminary meeting to agree overall strategy	________	________
4. Provincial programme coordinator selected	________	________
5. District coordinators selected	________	________
6. Executive officer selected and office allocated	________	________
7. Existing water supply inventories consolidated in the planning office with demographic data (urban and rural)	________	________
8. Basic data of inventories analysed levels of service, including coverage identified	________	________
9. Water-related disease data consolidated in the planning office	________	________
10. Existing (urban) surveillance activities summarized (number and type of analyses and inspections done each year in each water supply)	________	________
11. Existing quality control activities summarized (number and type of analyses done by water authority)	________	________
12. Assess compliance with WHO Guidelines in terms of quality of results and defined frequency of quality control sampling programme	________	________
13. Prepare inventories of health laboratory staff and sanitary technicians in province	________	________
14. Assess staff workload and identify suitable staff for surveillance	________	________
15. Prepare inventories of available health laboratory equipment and identify equipment needed and reagents required	________	________
16. Define proposed yearly surveillance activities as number of inspections and analyses at: provincial capital; district capitals; total community piped supplies; total unpiped water sources	________	________
17. Calculate and submit surveillance budget (salaries/training and operating costs/capitals costs)	________	________
18. Submit surveillance proposal 3–6 months ahead of intended start date	________	________
19. Reformulate priority surveillance activities in light of agreed budget constraints	________	________

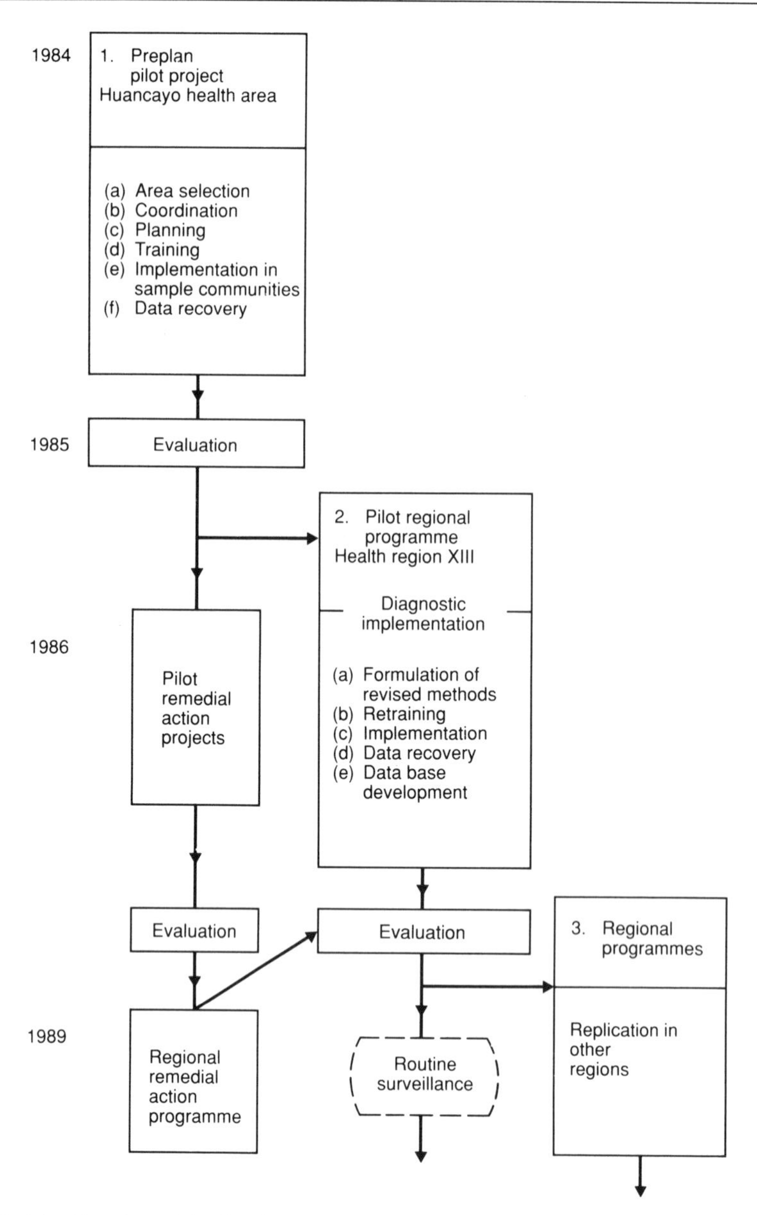

FIG. 4.1 Overview of the development strategy for extending water surveillance in Peru.

It may be noted that a national conference to launch the pilot projects is by no means essential and certainly does not guarantee the momentum to sustain the programme.

Peru's national conference launch will take place 5 years after the pilot projects' initiation. It would seem more important to identify and establish task

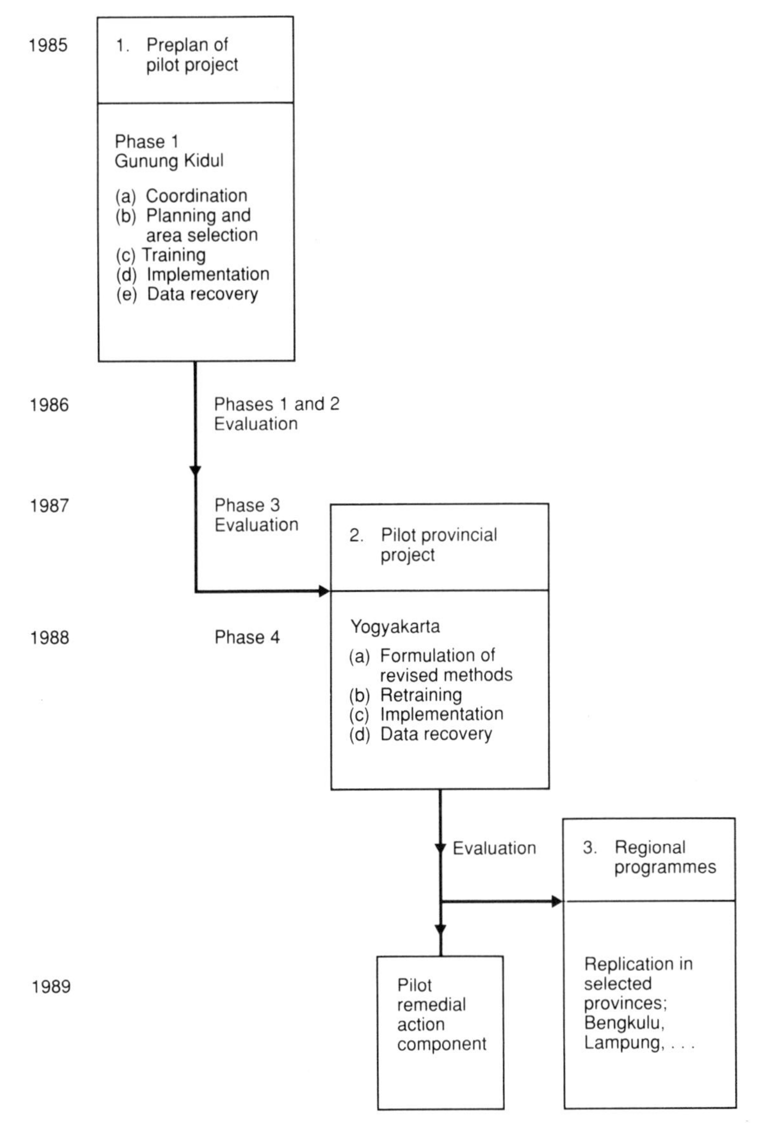

FIG. 4.2 Overview of the development strategy for extending water surveillance in Indonesia.

forces and action committees to initiate, guide and manage preplans and pilot projects. It is therefore recommended that a national conference launch should be held over until preliminary results are available. Encouraging results can then be used to convince senior politicians of the merit of a larger programme. Everybody likes to fund success stories, not failures! The main lesson is to start with small projects and sort out teething problems before constructing larger programmes.

4.2 Choice of methods

The basic methods adopted by all three projects at the start were those described in volume 3 of the WHO Guidelines (WHO 1984/85). This refers to the equal importance of sanitary inspection and water analysis; none the less all three projects were grossly deficient in sanitary inspection reporting, at least in the initial phase or preplan stage. This was partly attributable to the perception of project staff that the project was a water quality study, a view which is unfortunately reinforced by the title 'Drinking water *quality control* in small communities'. The deficiency in useful sanitary inspection data which are recorded in the preplan studies may also have been due to deficiencies in the preliminary survey training, but this training was also based on unsuitable record forms set out in volume 3. It is clear that the on-site sanitary survey is the most critical part of the practical methodology and we have therefore attempted a major revision which is presented here in Chapter 5.

Volume 3 of the WHO Guidelines gives at least twice as much space to basic water quality analysis as to inspection methods. A more balanced approach to inspection and analysis is to be found in *Surveillance of Drinking Water Quality* (WHO 1976a); none the less even this monograph has significant limitations with respect to the recommended sanitary surveys required by the pilot projects. As a consequence the projects have fulfilled a vital function in exposing those weaknesses and provided an opportunity to develop a broader conceptual and methodological approach to the whole subject of surveillance of drinking water.

Since surveillance is an investigative activity which is intended to identify and evaluate all conditions that pose a danger to health, all those factors which *indicate the quality of water supply service* should be included in reports. It should therefore be one of the major objectives of the surveillance agency periodically to summarize the critical indicators as part of its overall health risk assessment of water supply services. The Zambian and Indonesian projects limited their approach primarily to the assessment of bacteriological contamination, whereas the consultants for the Peruvian programme broadened the scope of its assessment to include the following *indicators of quality of service*:

Coverage	Percentage of the total population served with domestic connections *or* percentage with access to an improved source
Continuity	Hours/day and days/year that water was supplied
Quantity	Total volume/capita/day supplied or used
Quality	Classified primarily on faecal contamination
Cost	Tariff paid per month
Sanitary risk	Number of points of risk identified during inspection

4.3 Establishment of routine surveillance

Based upon the experience found with the three pilot projects a model methodology has been developed to facilitate the initiation of similar national projects in the future. This model is presented as a chronological series of procedural steps intended to identify the causes and sources of pollution and leading finally to protective and remedial action strategies. The principal elements are summarized in the form of a flow chart in Fig. 4.3.

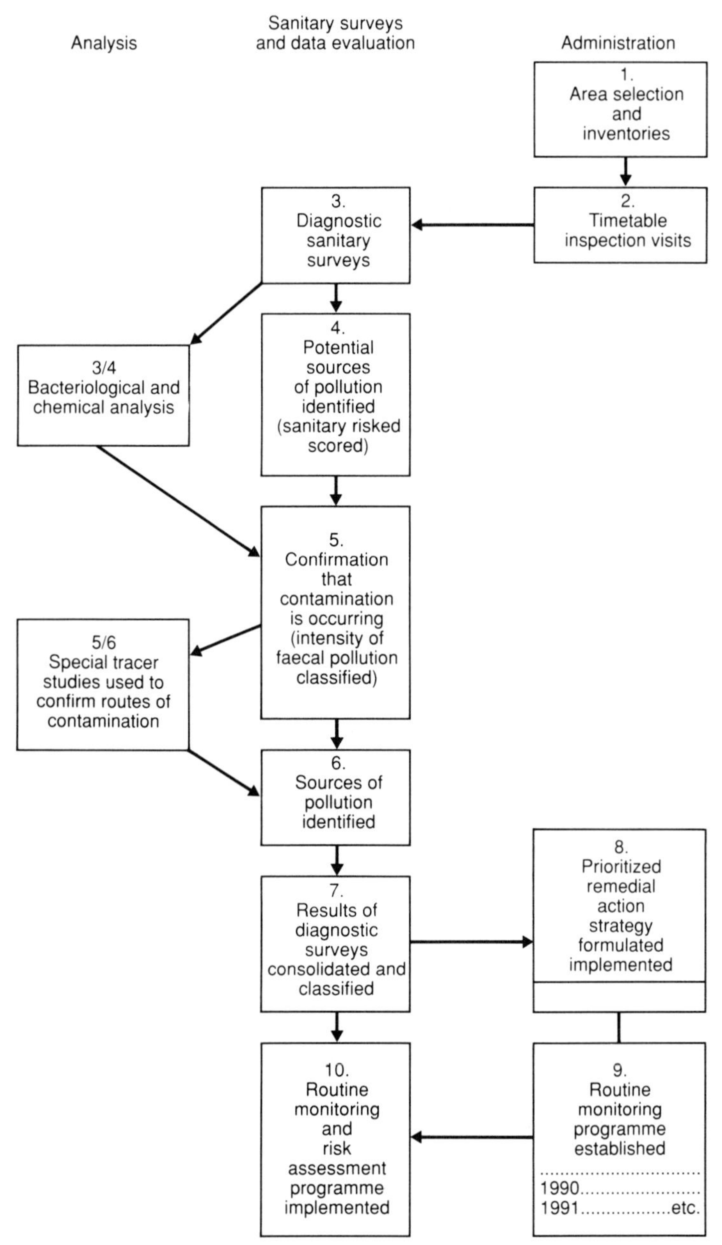

FIG. 4.3 Flow chart of procedural steps required for establishing a surveillance programme.

For the purpose of initiating a national programme in a pilot phase, the developmental phases shown in Fig. 4.1 may be followed. Experience with our pilot projects has shown that the logical way to implement routine surveillance services is on an incremental basis. The first stage is a diagnostic of a limited number of systems in a preplan project. Following evaluation of the preplan and revision of methods it may be expanded into a pilot project, again followed by evaluation. The next stage is expansion and replication throughout a department or region, and the last stage would be national replication. This is the approach which was adopted in Peru and Indonesia in the period 1985–90. The Peruvian project, in particular, demonstrated how the infrastructure evolves during the initial phases of programme development. Figure 4.4 shows the early and final stage of surveillance laboratory services. Examples are given in Figs 4.5 and 4.6.

The model for planning the development of laboratory services for surveillance followed in Indonesia and Peru was similar to that described in a WHO report, *Establishing and Equipping Water Laboratories in Developing Countries* (1986). This suggests a three-tier infrastructure:

1. *Central laboratory*. At the highest level there should be a central or national reference laboratory. This laboratory should be well staffed and equipped with conventional and advanced equipment capable, eventually, of analysing all WHO guideline parameters.
2. *Regional laboratories*. Located in provincial or regional capitals and staffed and equipped to analyse up to 35 parameters.
3. *Basic laboratories*. Located in smaller provincial towns or district capitals and equipped to analyse 5–10 basic parameters. Initially they should be equipped to analyse the WHO Guidelines volume 3 parameters: faecal coliform counts, turbidity and chlorine residual.

We would recommend that in the initial phase of development of surveillance that the third level should be based on field test kits, as shown in Fig. 4.4(a).

4.4 Conclusions

It is only worth while considering planning the development of surveillance programmes if the political will to implement water legislation exists. The legal instruments, if they exist, may be a powerful weapon which may be used to encourage government to initiate pilot projects and develop programmes. Many countries have based their water and health legislation on WHO guidelines.

Too often a pilot project of water quality surveillance survives only as long as external donor support is provided. The real challenge for water supply and surveillance agencies is to plan and make the transition from externally funded diagnostic pilot project to self-sustaining routine monitoring, control and maintenance programmes at national level. It requires enormous corporate political will and public service commitment to release and manage the rela-

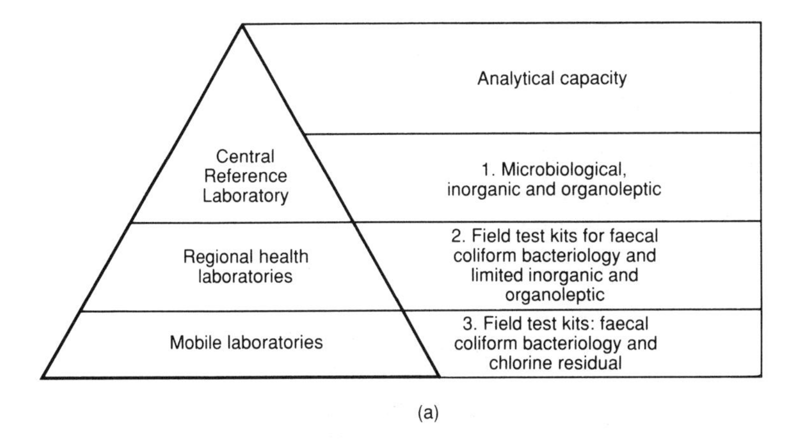

(a)

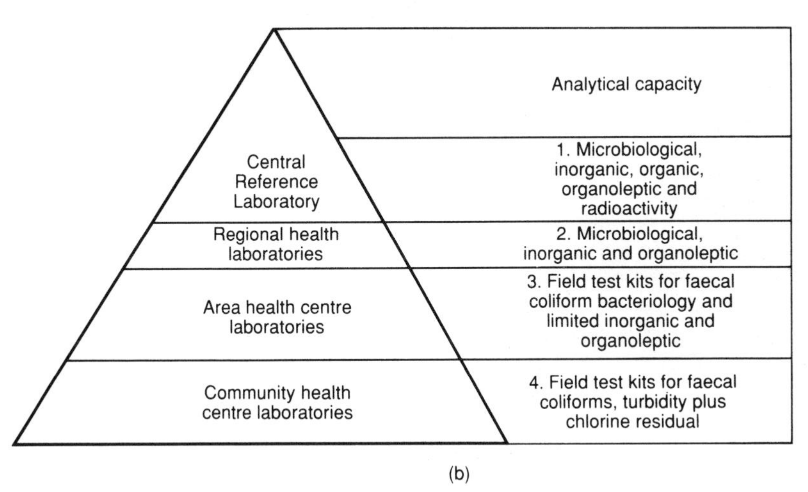

(b)

FIG. 4.4 Laboratory services infrastructure for water quality analysis: (a) initial structure; (b) final structure. (Lloyd *et al.* 1986b).

tively modest sums necessary to improve water quality services and protect water quality.

The development and continuity of routine surveillance are essential to the maintenance and improvement of health. This will only be sustained by staff prepared to plan ahead and struggle for the financial and human resources on which such public sector programmes are dependent. Therefore a vital function of the national surveillance agency is to provide training to the regional and provincial staff beyond the narrow technical scope to include management skills which provide a clear framework of surveillance. This type of planning and management training should unite urban and rural surveillance and include:

(a) institutional development and structure;
(b) communications strategies;

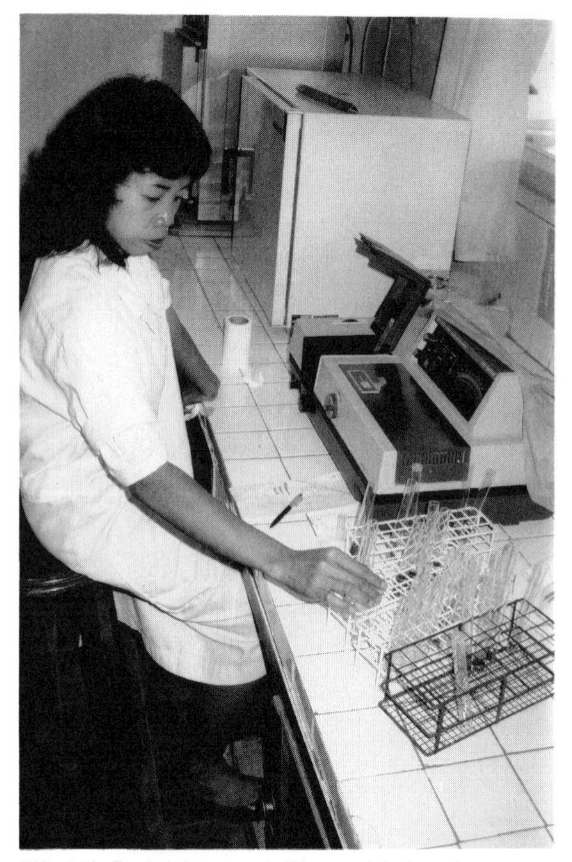

FIG. 4.5 Basic laboratory in Wonosari, Indonesia.

(c) resource planning and acquisition;
(d) team building;
(e) programme planning, promotion and implementation;
(f) phased planning and promotion of remedial action strategies.

Further information on this subject may be obtained from the Robens Institute, UK and DelAgua, Peru in a publication entitled *The National Surveillance Plan for Drinking Water Services*. This is available in English and Spanish.

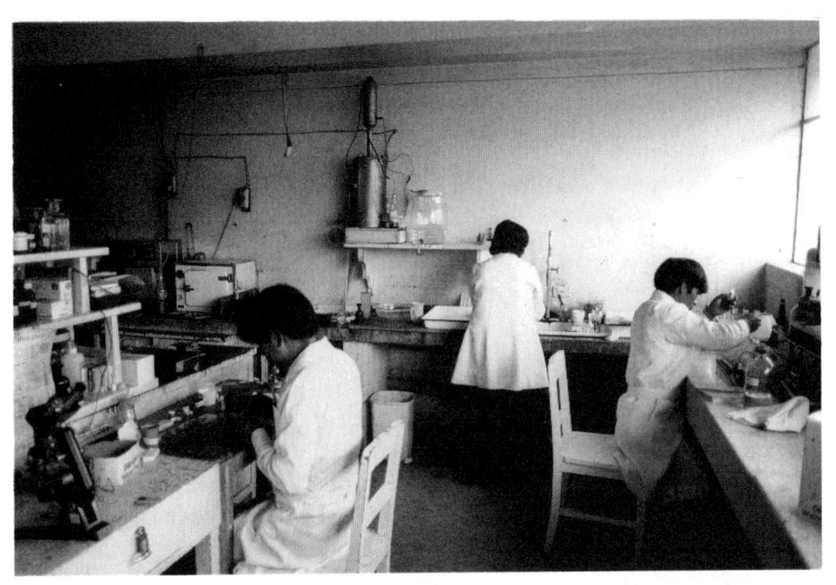

FIG. 4.6 Regional laboratory in Huancayo, Peru.

CHAPTER 5

Sanitary Inspections

The procedural steps for establishing a surveillance programme include several preliminary steps dealing with the development of a sound data base on the water supply sources in the project area. In accordance with the flow chart in Fig. 4.3, the first step is area selection and basic inventories. The present chapter will cover the steps up to the fourth one which leads to the identification of pollution problems which may threaten drinking water quality at the source or point of abstraction.

5.1 Inventories and inspection visits

The surveillance or environmental protection agency, at national level, will normally decide on the area selected for study on administrative and physiographic criteria. Generally, the most important administrative criterion is the size of the population served by a defined source (or sources) and this factor will tend to dominate the selection of the area.

The surveillance team, under the supervision of an area coordinator, should be charged with gathering basic data in order to prepare inventories in the area selected. The team may well require not only specific water surveillance training (as described in section 3.5), but also training for basic data gathering because they will be required to:

(a) record the population in each community and town;
(b) list the water sources which are being used;
(c) summarize the existing water quality data supplied by the water authority from the water quality archive;
(d) identify existing sources for which no data are available;
(e) consolidate water-related epidemiological data;
(f) match disease outbreaks and hot spots with water source and quality.

The last of the points listed above will generally be the most difficult to do because epidemiological reporting is based on political boundaries and these rarely correspond to water basins. In supplies where there are a number of water sources it will be necessary to map out and list the sources supplying each population zone.

The basic data should be validated independently wherever possible, by the national surveillance agency, by cross-checking, e.g. in the case of demographic data, with the most recent census. The first indicator of the success of a water supply development programme is *coverage*, and this should be calculated for each community and area before consideration is given to quality monitoring. To justify routine monitoring programmes the coverage should normally be above 50 per cent; however, in the case of new water source development projects prospective quality monitoring of surface water and groundwater following test drillings should be carried out when new sources are being considered for exploration.

In the early stages of a risk assessment and monitoring programme it will not be possible to make a definitive inventory of all sources and supplies; this will only be achieved by the end of the diagnostic phase. Thus for example in Table 5.1 the initial inventory from Ministry of Health records indicated that there were only 231 rural supplies in the health region, whereas by the end of the diagnostic, over 500 organized supplies had been entered into the inventory. Table 5.1 also shows the number of untreated gravity and pumped supply systems, but it cannot be assumed that these are from groundwater sources until a diagnostic sanitary survey has been carried out.

Table 5.1 Example summary of basic inventory of rural water supplies in the XIIIth health region built by Peruvian Ministry of Health, National Plan for Rural Water Supplies. Lloyd *et al.*, (1986a)

Health area no.	Population served	Coverage (%)	Type of system*			Total number of systems
			GST	GCT	BST	
47	14,459	63.03	14	3	0	17
48	65,665	51.49	83	10	7	100
49	34,716	56.04	50	3	1	54
50	19,303	52.75	28	6	1	34
51	11,208	64.26	15	6	0	21
52	2,869	49.1	4	0	0	4
Totals	148,200	54.4	194	28	9	231

* GST = Gravity without treatment
GCT = Gravity with treatment
BST = Pumped without treatment

In every country there are communities which are particularly at risk from water-related, including water-borne, diseases. It is impossible to prioritize investment to reduce the risk to these especially vulnerable groups, if reliable data on the risk to which they are exposed are not systematically collected. The inventory of populations and corresponding water services is the starting-point in the development of the essential data bases on which strategic interventions for improvement of sanitation services can be planned. This should be followed up by the second step (see Fig. 4.3), planning and timetabling sanitary inspection visits to all water supplies in order to assess the risk attributable to each source.

Recognizing the difficulties and costs involved in visiting sites and testing samples in remote communities it is important to have straightforward and pragmatic guidelines for timetabling visits. Inspection should be prioritized according to the following criteria:

(a) in the area selected the inventories should be as complete as possible;
(b) the largest urban populations' sources should be dealt with first;
(c) for rural supplies the largest number of people using a single source;
(d) for small rural supplies public facilities take precedence over private. Thus the overriding consideration will be the population served by each installation and this will dictate the frequency of visits and of bacteriological and chemical analysis.

Depending on the type and size of water supply systems, there are widely varying intervals between inspection visits proposed, ranging from 1 to 5 years. Table 5.2 shows that the inspection intervals are much longer than those recommended for bacteriological analysis.

Table 5.2 Suggested frequency of principal activities in surveys by surveillance agency

Population size served by source	Maximum interval between sanitary inspections	Maximum interval between bacteriological samples*
>100,000	1 year	1 day
50,000–100,000	1 year	4 days
20,000–50,000	3 years	2 weeks
5,000–20,000	3–5 years	1 month
<5,000		
Community dug wells	Initial, then as situation demands	As situation demands
Deep and shallow tubewells	Initial, then as situation demands	As situation demands
Springs and small borehole piped supplies	Initial and every 5 years, or as situation demands	As situation demands

* There are many countries where the water supply agency does not yet carry out routine bacteriological quality control even in urban areas; in such places there is a burden of responsibility on the surveillance agency to fulfil this role of analysis, with the frequencies indicated, until such time as the supply agency can take over the function. Only then may the surveillance agency reduce the frequency of analysis.

5.2 Diagnostic sanitary surveys

5.2.1 Improvement of survey procedures

Volume 3 of the WHO Guidelines (WHO 1984/85) refers to the equal importance and complementary nature of sanitary inspection and water analysis; none the less experience of the WHO pilot projects in Indonesia, Peru and Zambia in the period 1985–88 has demonstrated that the sanitary inspection report forms set out in volume 3 were unsuitable for determining sanitary risks attributable to rural water supplies. Volume 3 of the WHO Guidelines gives at least twice as much space to basic water quality analysis as to inspection methods. A more balanced approach to inspection and analysis is to be found in *Surveillance of Drinking Water Quality* (WHO 1976a); even this monograph has significant limitations with respect to the practical implementation of sanitary surveys and provides no guidance as to how to quantify or use risk information. It is clear that the on-site sanitary survey is a critical part of the practical methodology and the authors have therefore attempted a major revision here to try to investigate and demonstrate the relationships between sanitary risks and water quality.

The observable faults in the drinking water system, which may give rise to problems of quality or continuity of supply and hence to disease, can most readily be identified by careful on-site inspection. Every fault should be systematically listed during the sanitary survey and each point may therefore be considered as a *sanitary risk factor*.

Risk factors (such as construction of a latrine within 10 m of a community well, or the absence of an impervious concrete plinth around the well) are causes or signals which can be identified prior to the event that they predict, e.g. pollution of the well causing an outbreak of disease. Although it is possible that one risk factor may increase the chances of epidemic disease over and above others, we do not have sufficient information to weight and quantify this at present. On the other hand, it is logical that the greater the number of risk factors the greater the possibility that the community will be subjected to poor water quality and continuity, and as a consequence suffer from increased water-related disease. Every extra mechanical fault, or point of exposure to risk, may serve to increase the intensity of contamination and thus the risk to health. Similarly, every remedial action which eliminates a point of risk will reduce the probability of water-borne disease. Therefore, in the absence of information to the contrary:

Identifiable points of sanitary risk are weighted equally.

The equal weighting of risk factors is therefore the approach which has been adopted by the authors to develop a *sanitary risk score* based on sanitary inspection.

In the development of new supplies possible sources of contamination should be anticipated and avoided by proper source selection, design and construction practices. The main hazards which should be surveyed in new groundwater installations may be split between prospective surveys and those carried out during and on completion of an installation. Problem sites to be checked for are listed in Table 5.3 for the case of prospective investigation and during the initial phases of new sources.

Table 5.3 Principal hazards in groundwater installations

	Potential pollution source
Sanitary survey of prospective source	
Local caves, sink-holes or abandoned boreholes used for surface drainage or sewage disposal in the vicinity of the source, causing pollution of the groundwater table	Off site
Fissures or open faults in strata overlying water-bearing formations	Off and on site
Source located near sewers, pit latrines, cesspools, septic tanks, subsurface tile systems, drains, livestock concentrations (farmyards), pits below ground surface	Off site
Manufacturing, industrial or agricultural plant wastes discharged or spilled on watersheds or into underground strata causing contamination	Off site
Sanitary survey during construction and commissioning phase	
Casing of tubular wells leaking or not extended to a sufficient depth; absence of or inadequate sanitary seal around casing	On site
Casing of tubular wells not extended above ground or floor of pump room, or not closed at top, or casing used as suction pipe	On site
Leaks in system under vacuum	On site
Well heads, well casings, pumps, pumping machinery, exposed suction pipes, or valve boxes connected to suction pipes located in pits below the ground surface; hence potentially subjected to flooding	On site

5.2.2 Reporting of results

A sanitary survey has been designed for each of the main types of water sources and supplies listed in the project inventories. The objective is to establish a reporting system which can be rapidly and accurately completed on site at the same time that sampling and water quality analysis are carried out. The report form is intended to serve several purposes as follows:

1. identify all the potential sources of contamination of the supply;
2. quantify the level of risk of each groundwater facility;
3. provide a graphical means of explaining to the operators, or users in rural communities, the risks of contamination attributable to the facility (i.e. on-site hygiene education preparing for remedial action);
4. provide clear guidance for the user, and a record for the local surveillance

supervisor, as to the remedial action which is required (as shown in Figs 5.1–5.7).

To meet these needs, double page report forms have been designed and validated in pilot projects (as shown in Figs. 5.1–5.7). They cover the main types of rural installation starting with the most basic, thus:

1. dug wells;
2. converted covered dug wells with handpumps (or windlass);
3. deep and shallow tubewells with handpumps;
4. boreholes with mechanized pumping;
5. gravity-fed protected spring sources;
6. simple distribution systems;
7. rainwater catchment.

It is important to note that although the survey forms presented in the following pages may be used as a model pilot methodology for risk assessment the authors *do not consider them to be the definitive statement in contamination risk assessment. They should be extended or modified in the light of local needs.*

On site the sanitarian should first complete the checklist on the report form with the assistance of the operator, community representatives or owner. The number of risk points can be immediately totalled to give a sanitary inspection risk score in the *range 0 (no risk) to 10 or more (very high risk).*

After completing the checklist the sanitarian should circle each of the points of risk on the diagram, preferably in red ink. The diagram should be separated from the report form and given to the owner or community representative together with instructions and explanation of what needs to be done to improve the facility. The recipient should sign the report form which the sanitarian retains for the surveillance centre records.

The sanitarian carrying out the sanitary survey should record whether or not sampling or analysis is or will be undertaken (step 3/4 in Fig. 4.3). In some programmes it has been shown to be most labour saving and hence time saving to carry out the analysis in the field at the same time as the inspection is done; elsewhere water analysis may be done as a follow-up.

5.3 Pollution source identification

In the Peruvian inventory, summarized in Table 5.1, 100 per cent of pumped systems (BST) were derived from groundwater and the majority (74 per cent) of simple gravity systems (GST) have protected springs as the source, but this was only revealed by the diagnostic sanitary surveys. The remaining 26 per cent of simple gravity systems have either unprotected spring sources or surface-water sources; they may thus be divided into three groups and distinguished in the inventory as follows [on p. 82]:

1. Springs, where the eye is fully enclosed and protected (capped).
2. Springs, where the intake is open and the source is collected in unprotected

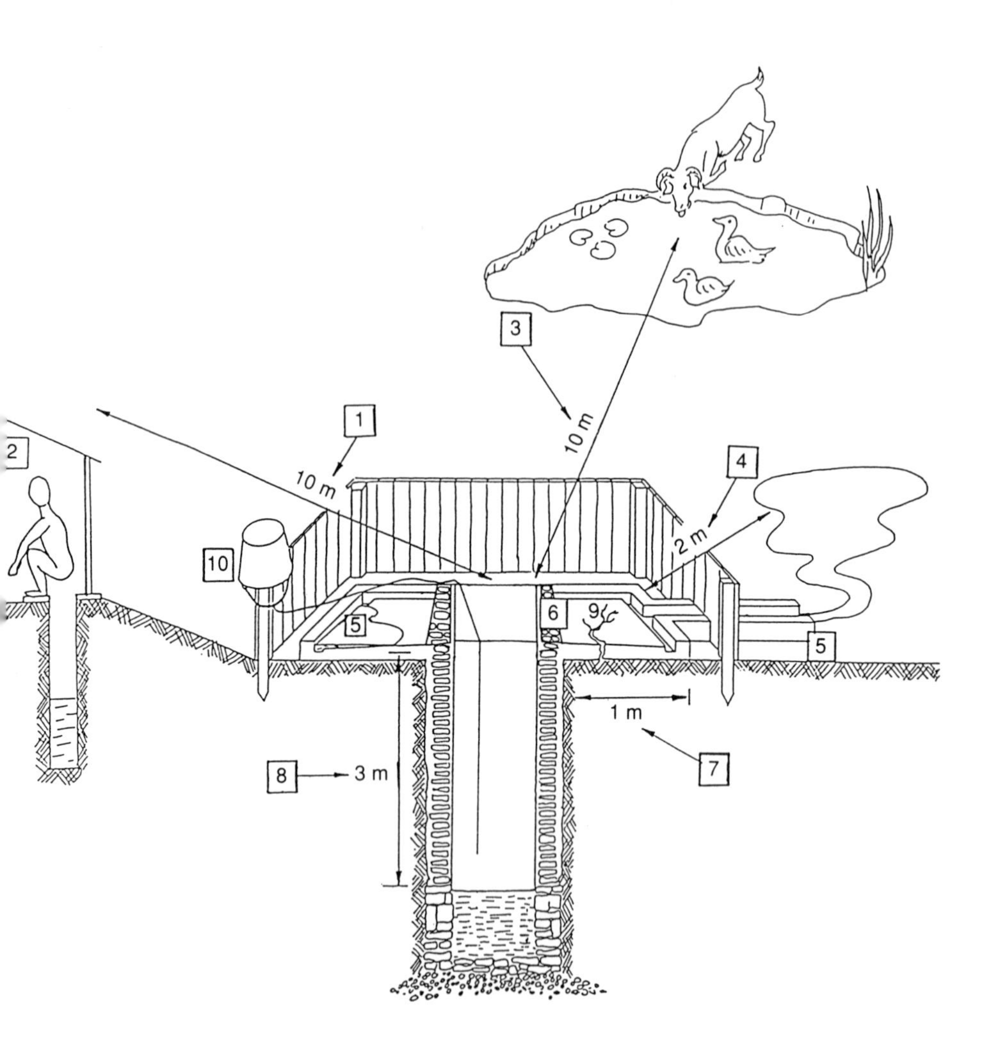

Circle all sanitary risks in red.

Advise pot chlorination where risk score is greater than 3.

Separate sheet along dotted line & give the diagram to the owner of the facility.

FIG. 5.1 Dug well sanitary survey.

WATER SURVEILLANCE AND IMPROVEMENT PROGRAMME

SANITARY SURVEY FORM FOR THE ASSESSMENT OF RISKS
OF CONTAMINATION OF DRINKING WATER SOURCES

I **Type of facility** **DUGWELL**

General information :

1 Location : Health Centre ....

: Village ....

2. Code No ..../........../..........

3. Water authority/Community Representative signature ....

4. Date of visit ....

5. Is water sample taken?..........Sample No .... Faecal coliform grade..........

II **Specific Diagnostic Information for Assessment**

		Risk Yes	Risk No
1.	Is there a latrine within 10m of the well?	☐	☐
2.	Is the nearest latrine on higher ground than the well?	☐	☐
3.	Is there any other source ofpollution within 10m of the well? (e.gs. animal excreta, rubbish)	☐	☐
4.	Is the drainage poor causing stagnant water within 2m of well?	☐	☐
5.	Is there a faulty drainage channel? Is it broken, permitting ponding?	☐	☐
6.	Is there an inadequate wall (parapet) around the well which would allow surface water to enter the well?	☐	☐
7.	Is the cement floor less than 1m wide round the well?	☐	☐
8.	Are the walls of the well inadequately sealed at any point for 3m below ground?	☐	☐
9.	Are there any cracks on the cement floor around the well which could permit water to enter the well?	☐	☐
10.	Are the rope and bucket left in such a position that they may be contaminated?	☐	☐

Total score of risks ..../10

Contamination risk score: 9/10 = V. high; 6 - 8 = high; 3 - 5 = intermediate; 0 - 2 = low

III. **Results and recommendations:**

(list nos 1 - 10)

The following important points of risk were noted:
and the authority advised on remedial action

Signature of sanitarian ....

FIG. 5.1 *cont* Dug well sanitary survey.

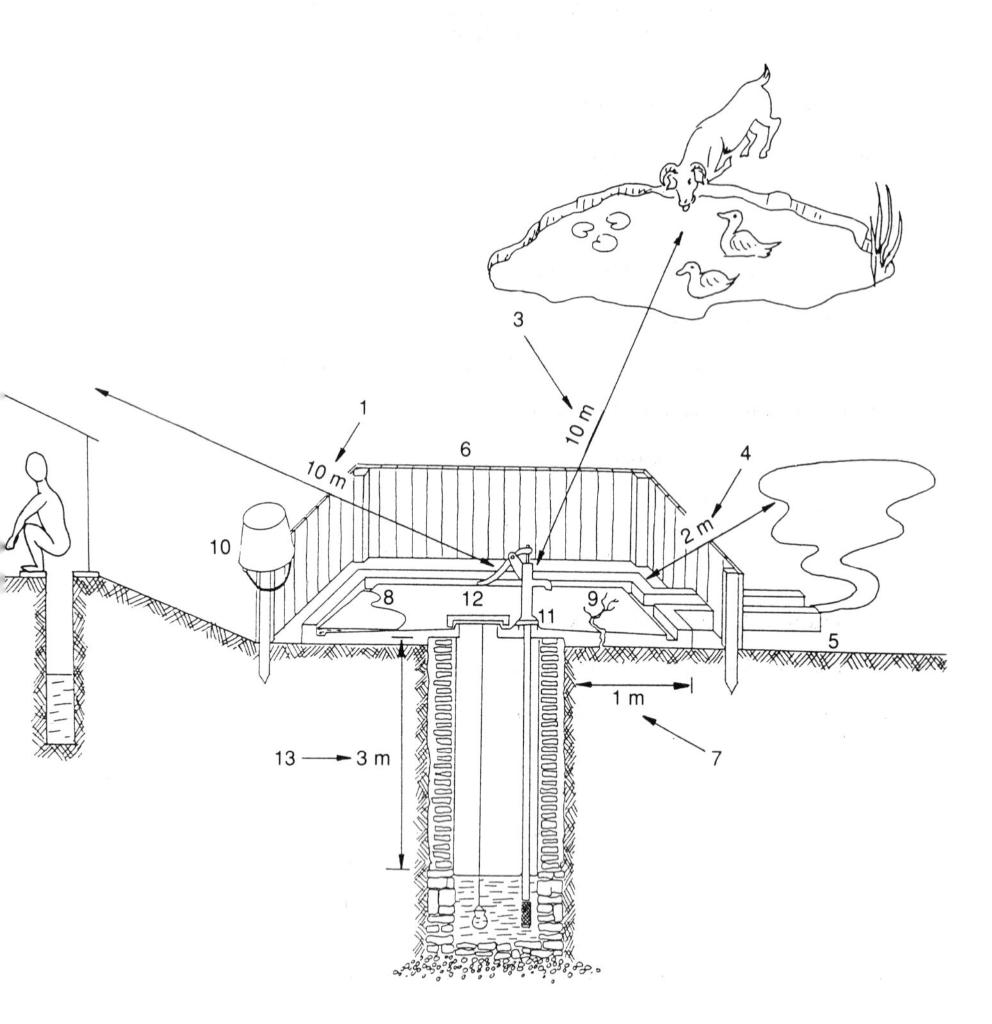

Circle all sanitary risks in red.
Advise pot chlorination where risk score is greater than 3.

Separate sheet along dotted line & give the diagram to the owner of the facility or the community representative.

FIG. 5.2 Converted handpumped dug well sanitary survey.

WATER SURVEILLANCE AND IMPROVEMENT PROGRAMME

SANITARY SURVEY FORM FOR THE ASSESSMENT OF RISKS
OF CONTAMINATION OF DRINKING WATER SOURCES

I Type of facility **HANDPUMP ON DUGWELL**

General information :

1 Location : Health Centre ....

: Village ....

2. Code No ..../../..

3. Water Authority/Community Representative signature ....

4. Date of visit ....

5. Is water sample taken?.................................Sample No .... Faecal coliform grade...........................

II Specific Diagnostic Information for Assessment

		Risk Yes	Risk No
1.	Is there a latrine within 10m of handpump?	☐	☐
2.	Is the nearest latrine on higher ground than the handpump?	☐	☐
3.	Is there any other source of pollution within 10m of the handpump? (e.gs. animal excreta, rubbish, surface water)	☐	☐
4.	Is there any ponding of stagnant water within 2m of the cement floor of handpump?	☐	☐
5.	Is the handpump drainage channel faulty e.gs, is it broken, permitting ponding? Does it need cleaning?	☐	☐
6.	Is there inadequate fencing around the installation, which would allow animals in?	☐	☐
7.	Is the cement floor less than 1m radius all round the handpump?	☐	☐
8.	Is there any ponding on the cement floor around the handpump?	☐	☐
9.	Are there any cracks on the cement floor around the handpump?	☐	☐
10.	Is a bucket also in use and left in a place where it could be contaminated?	☐	☐
11.	Is the handpump loose at the point of attachment to base? (which could permit water to enter the casing)	☐	☐
12.	Is the cover of the well insanitary?	☐	☐
13.	Are the walls of the well inadequately sealed at any point for 3m below ground level?	☐	☐

Total score of risks ..../13

Contamination risk score: 9/13 = V. high; 6 - 8 = high; 3 - 5 = intermediate; 0 - 2 = low

III. Results and recommendations: (list nos 1 - 13)

The following important points of risk were noted: ☐ ☐ ☐ ☐ ☐
and the authority advised on remedial action

Signature of sanitarian ....

FIG. 5.2 *cont* Converted handpumped dug well sanitary survey.

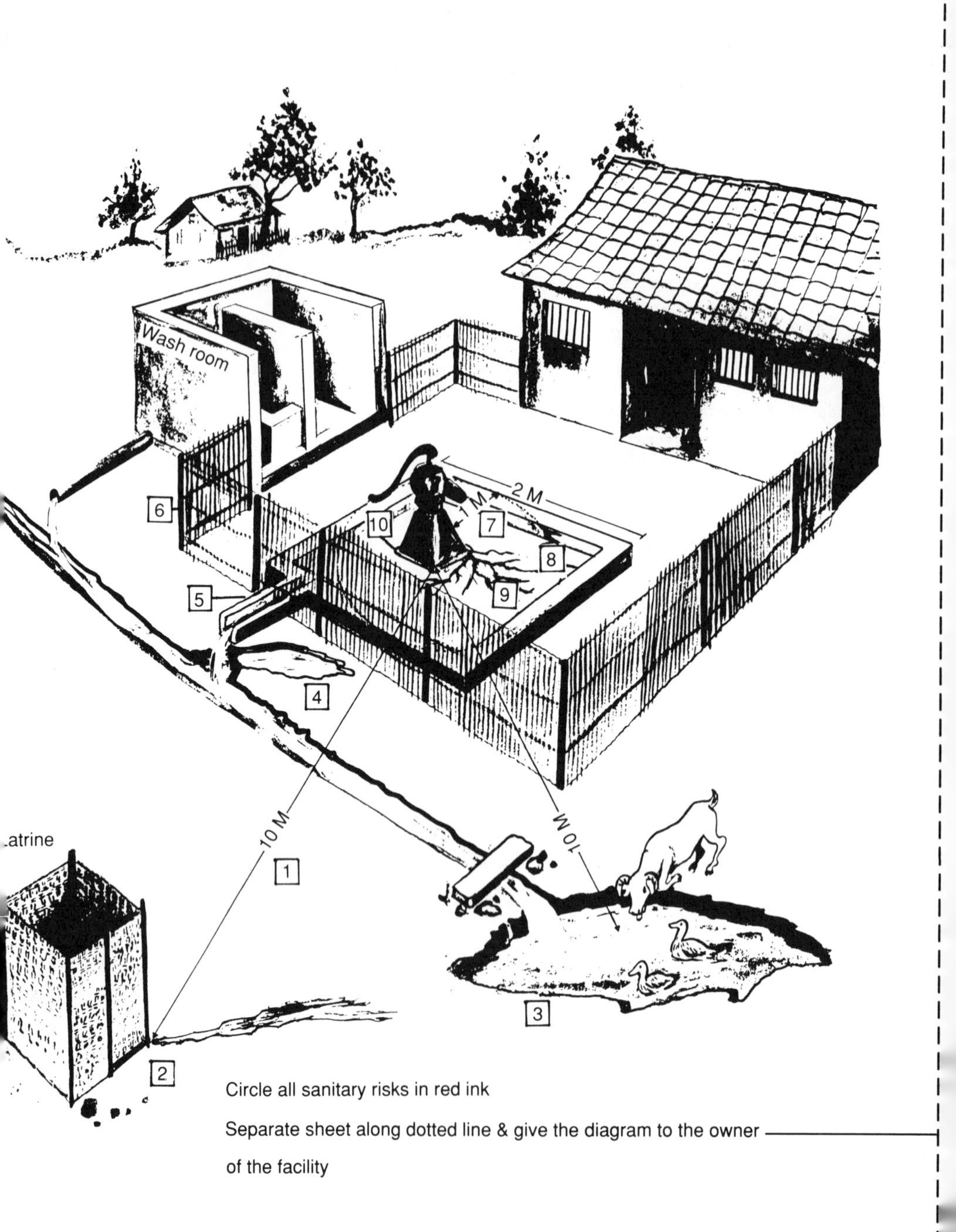

FIG. 5.3 Shallow and deep handpumped tubewells.

WATER SURVEILLANCE AND IMPROVEMENT PROGRAMME

SANITARY SURVEY FORM FOR THE ASSESSMENT OF RISKS
OF CONTAMINATION OF DRINKING WATER SOURCES

I **Type of facility** **SHALLOW AND DEEP HANDPUMPS (TUBEWELL)**

General information :

1 Location : Health Centre ....

: Village ....

2. Code No ..../../...

3. Water Authority/Community Representative signature ....

4. Date of visit ....

5. Is water sample taken?..................................Sample No .... Faecal coliform grade.............................

II **Specific Diagnostic Information for Assessment**

		Risk Yes	No
1.	Is there a latrine within 10m of handpump?	☐	☐
2.	Is the nearest latrine on higher ground than the handpump? (a pit latrine that percolates to soil)	☐	☐
3.	Is there any other source of pollution within 10m of the handpump? (e.gs. animal excreta, rubbish, surface water)	☐	☐
4.	Is there any ponding of stagnant water within 2m of the cement floor of handpump?	☐	☐
5.	Is the handpump drainage channel faulty? Is it broken, permitting ponding? Does it need cleaning?	☐	☐
6.	Is there inadequate fencing around the installation which would allow animals in?	☐	☐
7.	Is the cement floor less than 1m radius all round the handpump?	☐	☐
8.	Is there any ponding on the cement floor around the handpump?	☐	☐
9.	Are there any cracks on the cement floor around the handpump?	☐	☐
10.	Is the handpump loose at the point of attachment to the base? (which could permit water to enter the casing)	☐	☐

Total score of risks ..../10

Contamination risk score: 9/10 = V. high; 6 - 8 = high; 3 - 5 = intermediate; 0 - 2 = low

III. **Results and recommendations:** (list nos 1 - 10)

The following important points of risk were noted: ☐ ☐ ☐ ☐ ☐
and the authority advised on remedial action

Signature of sanitarian ....

FIG. 5.3 *cont* Shallow and deep handpumped tubewells.

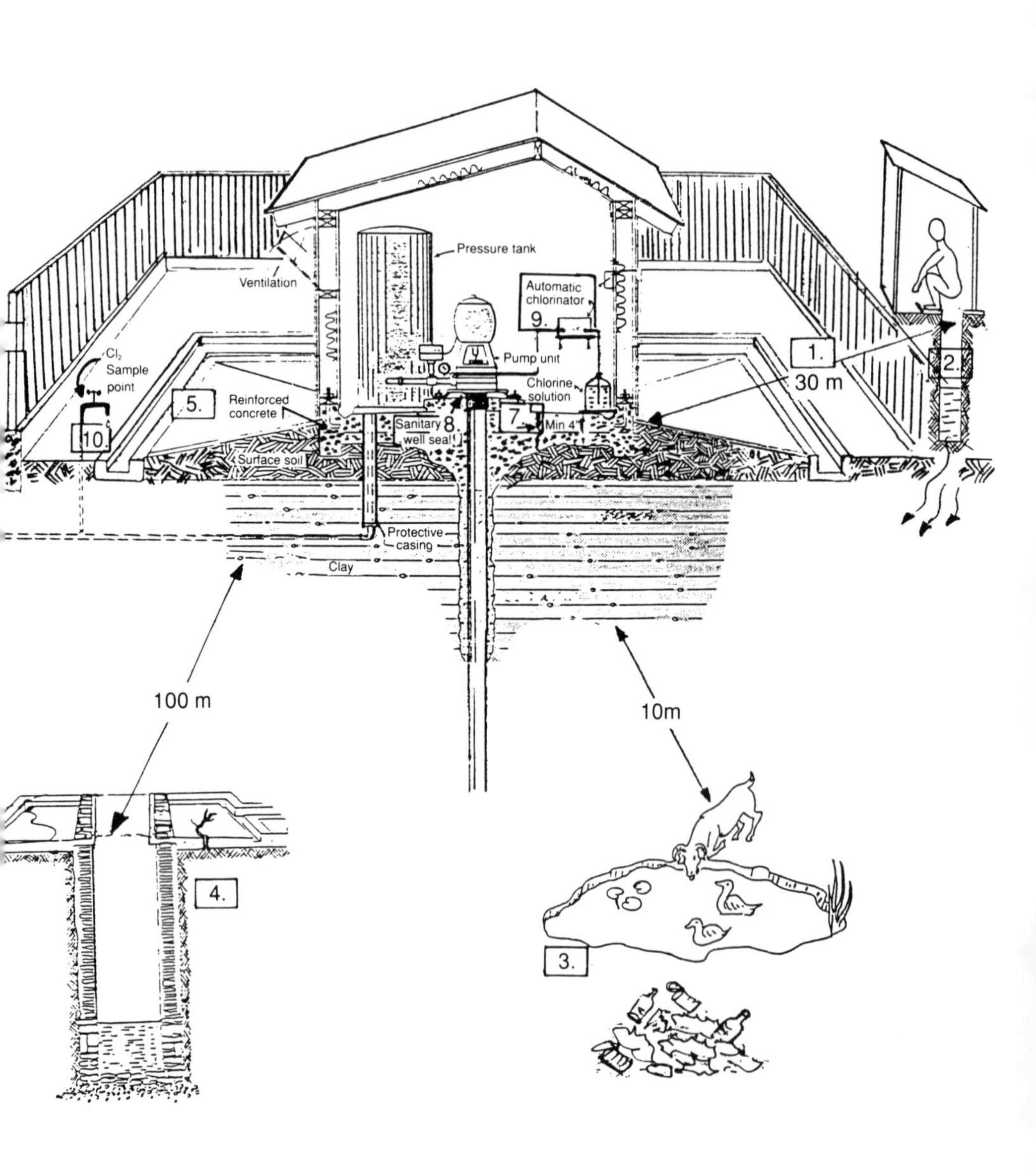

FIG. 5.4 Deep borehole with mechanical pumping.

WATER SURVEILLANCE AND IMPROVEMENT PROGRAMME

SANITARY SURVEY FORM FOR THE ASSESSMENT OF RISKS
OF CONTAMINATION OF DRINKING WATER SOURCES

I **Type of facility** **DEEP BOREHOLE**
General information :
1 Location : Health Centre ....
: Village ....

2. Code No ..../........../..........

3. Water Authority .... Signature ....

4. Date of visit ....

5. Is water sample taken?..........Sample No ....Faecal coliform grade ....

II **Specific Diagnostic Information for Assessment**

		Risk Yes	Risk No
1.	Is there a latrine or sewer within 30m of the pumphouse?	☐	☐
2.	Is the nearest latrine unsewered? (a pit latrine that percolates to soil)	☐	☐
3.	Is there any other source of pollution within 10m of the well? (e.g. surface water, animal excreta, rubbish)	☐	☐
4.	Is there an uncapped well within 100m of the borehole?	☐	☐
5.	Is the drainage area around the pumphouse faulty? (permitting ponding and/or leakage to ground)	☐	☐
6.	Is the fencing around the installation damaged in any way which would allow animals access or any unauthorised entry?	☐	☐
7.	Is the floor of the pumphouse permeable to water?	☐	☐
8.	Is the well seal insanitary?	☐	☐
9.	Does the chlorination record show any interruption in dosing? (if there is no record of chlorination, risk (yes) should be recorded)	☐	☐
10.	Is the free chlorine residual at the sample tap less than 0.2mg/l?	☐	☐

Total score of risks ..../10

Contamination risk score: 9/10 = V. high; 6 - 8 = high; 3 - 5 = intermediate; 0 - 2 = low

III. **Results and recommendations:**

The following important points of risk were noted: (list nos 1 - 10) ☐ ☐ ☐ ☐ ☐ ☐ ☐
and the authority advised on remedial action

Signature of sanitarian ....

FIG. 5.4 *cont* Deep borehole with mechanical pumping.

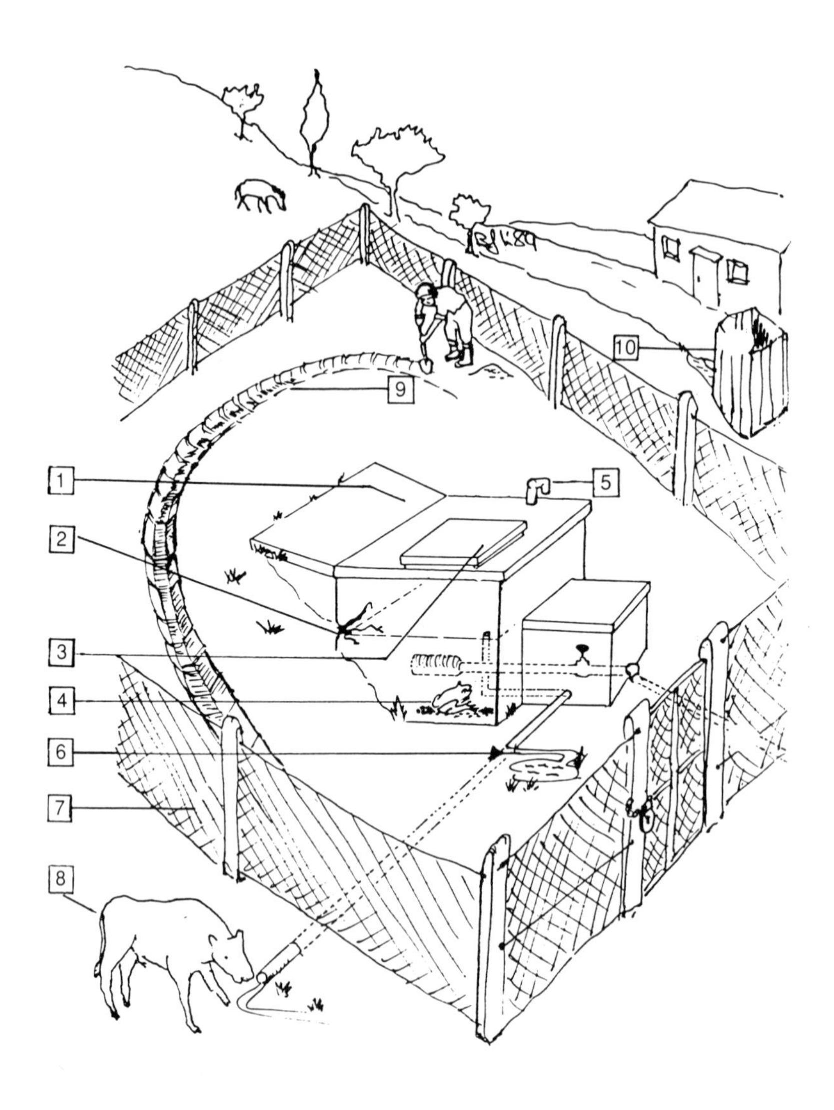

FIG. 5.5 Protected spring sanitary survey.

WATER SURVEILLANCE AND IMPROVEMENT PROGRAMME

SANITARY SURVEY FORM FOR THE ASSESSMENT OF RISKS
OF CONTAMINATION OF DRINKING WATER SOURCES

I Type of facility **GRAVITY FEED PIPED SPRING WATER SYSTEM**

General information :

1 Location : Health Centre ....

: Village ....

2. Code No ..../................/................

3. Water Authority .... Signature ....

4. Date of visit ....

5. Is water sample taken?................Sample No .... Faecal coliform grade ....

II Specific Diagnostic Information for Assessment

		Risk Yes	Risk No
1.	Is the spring source unprotected by masonry or concrete wall or spring box (open to surface contamination)?	☐	☐
2.	Is the masonry protecting the spring source faulty?	☐	☐
3.	If there is a spring box, is there an insanitary inspection cover in the masonry?	☐	☐
4.	Does the spring box contain contaminating silt or animals?	☐	☐
5.	If there is an air vent in the masonry, is it insanitary?	☐	☐
6.	If there is an overflow pipe, is it insanitary?	☐	☐
7.	Is the area around the spring unfenced?	☐	☐
8.	Can animals have access within 10m of the spring source?	☐	☐
9.	Is the spring lacking a surface water diversion ditch above it, or (if present) is it non functional?	☐	☐
10.	Is there any latrine upstream of the spring?	☐	☐

Total score of risks ..../10

Contamination risk score: 9/10 = V. high; 6 - 8 = high; 3 - 5 = intermediate; 0 - 2 = low

III. Results and recommendations: (list nos 1 - 10)

The following important points of risk were noted: ☐☐☐☐☐☐☐
and the authority advised on remedial action

Signature of sanitarian ....

FIG. 5.5 *cont* Protected spring sanitary survey.

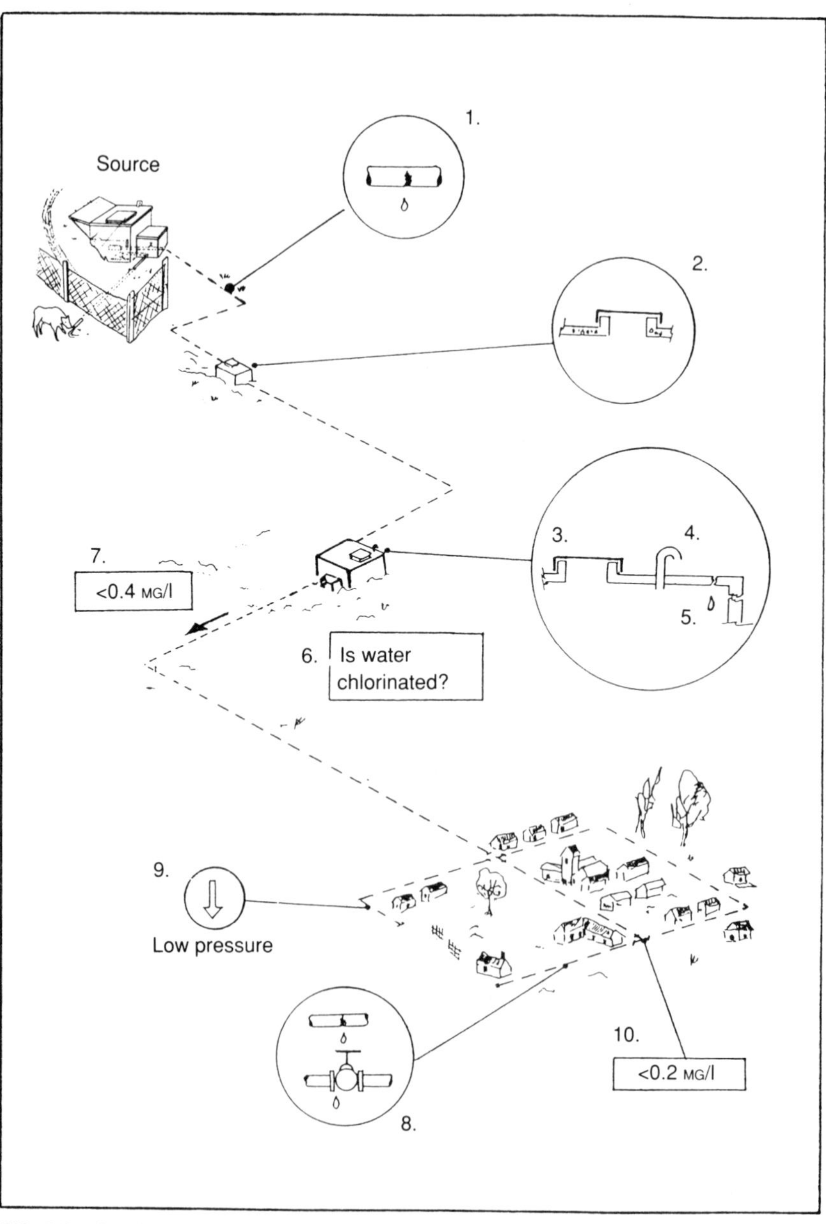

FIG. 5.6 Gravity feed piped distribution system sanitary survey.

WATER SURVEILLANCE AND IMPROVEMENT PROGRAMME

SANITARY SURVEY FORM FOR THE ASSESSMENT OF RISKS
OF CONTAMINATION OF DRINKING WATER SOURCES

I Type of facility **GRAVITY FEED PIPED SUPPLIES**

General information :

1 Location : Health Centre ....

: Village ....

2. Code No ..../.../...

3. Water authority/Community Representative signature ....

4. Date of visit ....

5. Is water sample taken?...................................Sample No .... Faecal coliform grade.............................

II Specific Diagnostic Information for Assessment

	Risk Yes	Risk No
Conduction pipe to reservoir		
1. Is there any point of pipe leakage between the source and the reservoir?	☐	☐
2. If there are any pressure break boxes, are their covers insanitary?	☐	☐
3. Is the inspection cover on the reservoir insanitary?	☐	☐
4. Are any air vents insanitary?	☐	☐
5. Do the roof and walls of the reservoir allow any water to enter (is the reservoir cracked?)	☐	☐
6. Is the reservoir water unchlorinated?	☐	☐
Distribution pipes		
7. Does the water entering the distribution pipes have less than 0.4 ppm free residual chlorine (<0.4 mg/l)?	☐	☐
8. Are there any leaks in any part of the distribution system?	☐	☐
9. Is pressure low in any part of the distribution system?	☐	☐
10. Does any sample of water in the principal distribution pipes have less than 0.2 ppm free residual chlorine?	☐	☐

Total score of risks ..../10

Contamination risk score: 9/10 = V. high; 6 - 8 = high; 3 - 5 = intermediate; 0 - 2 = low

III. Results and recommendations: (list nos 1 - 10)

The following important points of risk were noted: ☐☐☐☐☐
conduction pipe to the reservoir ....
the distribution system ....
and the authority advised on remedial action

Signature of sanitarian ....

FIG. 5.6 *cont* Gravity feed piped distribution system sanitary survey.

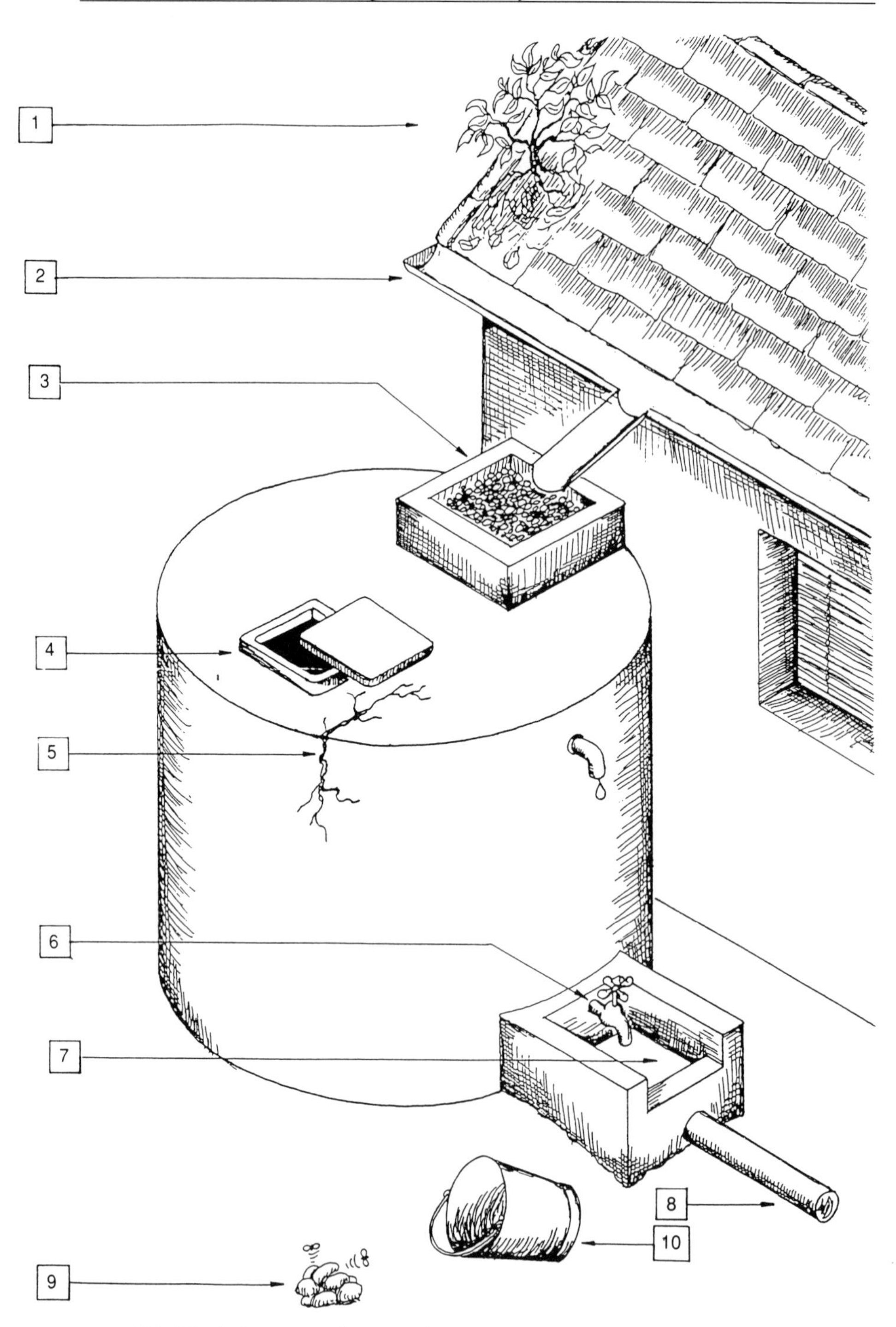

FIG. 5.7 Rainwater catchment tank sanitary survey.

WATER SURVEILLANCE AND IMPROVEMENT PROGRAMME

SANITARY SURVEY FORM FOR THE ASSESSMENT OF RISKS OF CONTAMINATION OF DRINKING WATER SOURCES

0. Type of facility : RAINWATER TANK CATCHMENT

I. 1. General information :
   Location : Health Centre ....
   village ....
   2. Code no. ..../ ..../ ..../ ..../
   3. Owner of facility : .... Signature ....
   4. Date of visit : ....
   5. Is water sample taken? .... Sample no. ....

II. Specific Diagnostic Information

	RISK YES	RISK NO
1. Is there any visible contamination of the roof catchment area? (e.gs. plants, dirt or excreta)	☐	☐
2. Are the guttering channels which collect water dirty?	☐	☐
3. Is there any deficiency in the filter box at the tank inlet? (e.gs. lacks fine gravel)	☐	☐
4. Is there any other point of entry to the tank which is not properly covered?	☐	☐
5. Is there any defect in the walls or top of the tank (e.g. cracks) which could let water in?	☐	☐
6. Is the tap leaking or other wise defective?	☐	☐
7. Is the cement floor under the tap defective or dirty?	☐	☐
8. Is the water collection area inadequately drained?	☐	☐
9. Is there any source of pollution around the tank or water collection area? e.g. excreta.	☐	☐
10. Is the water bucket left in such a position that it may be contaminated?	☐	☐
Total Score of risks	____/10	

Contamination Risk Score 9-10 = V.high. 6-8 = High. 3-5 = Intermediate. 0-2=Low.

III. Results and Recommendations :

The following important points of risk were noted: and the owner advised on remedial action

(List nos.1-10) ☐☐☐☐☐☐

Signature of sanitarian ....

FIG. 5.7 *cont* Rainwater catchment tank sanitary survey.

channels and therefore subject to contamination, and consequently always at risk of contamination.
3. Raised surface-water sources, which are therefore always subject to gross contamination. These are therefore invariably high risk, and dealt with separately.

Water quality problems may be associated with, and traceable to, any or all of the following:

(a) poor quality source waters;
(b) poor site selection or protection;
(c) construction difficulties; and
(d) structural deterioration with age.

All of the problems above should be identifiable by the comprehensive sanitary survey; whether or not they are detected will depend largely on the thoroughness and perspicacity of the sanitarian. The sanitarian should always be on the look-out for additional points of risk, not just those listed in the survey report form!

Using the data derived from the sanitary surveys of point source systems each type of facility can be reviewed to assess, in rank order, the commonest points of risk. This approach was first applied in Java and Indonesia where the most common problems identified were as listed in Table 8.1. This table serves to stress how common many of the risk factors are and hence how important basic structural repairs are in the protection of the source water. A combination of factors, such as ponding around or on cracked plinths together with permeable well linings will inevitably lead eventually to pollution of the source.

It is obvious that open dug wells are far more vulnerable to contamination than, for example, tubewells, since dug wells are exposed to direct contamination from whatever falls into the well. None the less dug wells remain a principal source of rural water in many countries and are therefore likely to be the focus for low-cost improvement strategies as described in step 8 (see Fig. 4.3). At this stage, however, it is appropriate to point out that in Indonesia the most frequent fault common to open and covered dug wells, and to tubewells, is the unsuitability of drainage pipes (see also Table 8.1). These are typically too short, resulting in ponding of water within several metres of the well head. This in turn attracts animals and aggravates the risk of faecal contamination of water which will seep into the well.

It can thus be seen that potential risks exist from a defined number of sources and that analysis is now required to determine whether it is occurring and what the intensity of contamination is.

5.4 Conclusions

The methodology proposed in this chapter can only be improved with more field experience. It is therefore presented to surveillance and quality control programme managers, and especially to rural surveillance supply project staff, as a working hypothesis for field evaluation in their project areas. There are inadequacies in the system proposed; in particular the equal weighting of different points of risk can be criticized and important additional factors may be added in due course. Operation and maintenance, of handpumps for example, have been excluded although it is obvious that breakdowns impose a major health risk. Another factor is the importance of the radius of the concrete plinth protecting wells and the safe distance of the nearest latrine from the well point. Examples from the pilot projects are illustrated in Figs 5.8 and 5.9.

The complementary nature of sanitary inspection and analysis cannot be overemphasized. There are many occasions when the source of contamination is not visible by sanitary inspection. In the case of groundwater contamination, for example, a handpumped tubewell may appear to be in a good physical state, but the aquifer itself may be contaminated at a point remote from the tubewell facility. Thus remote contamination of the aquifer can only be detected by bacteriological or physico-chemical analysis. On the other hand, a single water sample is only representative of the moment in time when that sample is taken and changing environmental conditions, particularly rainfall, may quickly alter the level of contamination of a poorly protected source. Thus the sanitary inspection should at the very least reveal the most obvious points of contamination risk and, as has been demonstrated, can provide a robust and conservatively safe method of risk identification. It would be most unsatisfactory if sanitary inspection routinely underestimated bacteriological contamination. Happily the data presented in Chapter 7 show that this is rarely the case and that generally the sanitary inspection reveals more of the chronic risks of contamination than can be revealed by a single and costly bacteriological examination. This is not an argument for dispensing with bacteriological testing but rather for an economical and intelligent approach to bacteriological testing where funding is limited. The sanitary inspection methodology described above is primarily concerned with the identification of the commonest risks associated with structural defects in the water supply facility. Geographically, we have also focused on the identification of risks associated with contamination in the immediate vicinity of the source. We have given little attention to the assessment of the entire catchment area from which a source is derived. In the context of surface water sources, this is an important problem but has yet to be systematically developed. For groundwater, the surveillance of aquifer vulnerability is being effectively developed under a collaborative regional groundwater programme between the Hydrogeology Division of the British Geological Survey and CEPIS in Latin America. A series of reports on this complementary risk assessment approach is available from CEPIS in Lima.

FIG. 5.8 Sanitarian carrying out sampling and inspection of tubewell installation in Java. Note cracked plinth.

FIG. 5.9 Shallow well with windlass (Mongu district, Zambia). Note that although drainage channel is not working and plinth is cracked, water quality was protected by concrete rings below ground. (Photo: S. Sutton).

CHAPTER 6

Bacteriological and Physico-chemical Analysis

Whereas potential sources of pollution may appear to be obvious and can be identified during sanitary surveys, actual contamination can only be assessed by analysis. It is worth emphasizing that the two activities are complementary and may usefully be carried out at the same time. Sometimes, however, either due to poor time management or lack of resources, analysis may be done as a follow-up activity.

It is most cost-effective for the sanitarian to make available at least a half-day per week for the surveillance activity so that at least four inspections and samples can be done in the time allocated. With experience, patterns will quickly emerge in which it will be obvious, in some cases, that the sanitary inspection is so bad that pollution is inevitable. In poorly funded rural programmes it may not always be possible or even necessary to carry out analysis on every source. In this case analysis should be reserved for the community sources serving most people.

6.1 Choice of parameters

As a caveat against excessive economy measures it should, however, be noted as an axiom that it is *far more important to examine a supply frequently using simple, reliable tests rather than less frequently with a more complex series of tests.* For this reason the following minimum critical parameter approach is recommended in volume 3 of the WHO Guidelines (WHO 1984/85).

A minimum of three critical and complementary parameters are essential for the microbiological assessment of the hygienic quality of drinking water, which are:

	Guideline value
1. faecal coliform count;	0/100 ml
2. chlorine residual;	0.2–0.6 mg/h
3. turbidity.	<1 TU and never >5 TU

There is a strong justification for applying the faecal coliform test as the only bacteriological test in basic monitoring programmes. The presence of any faecal coliform bacteria is of concern and the WHO guideline value is zero/100 ml in drinking water. Groundwater often meets this standard but in areas where

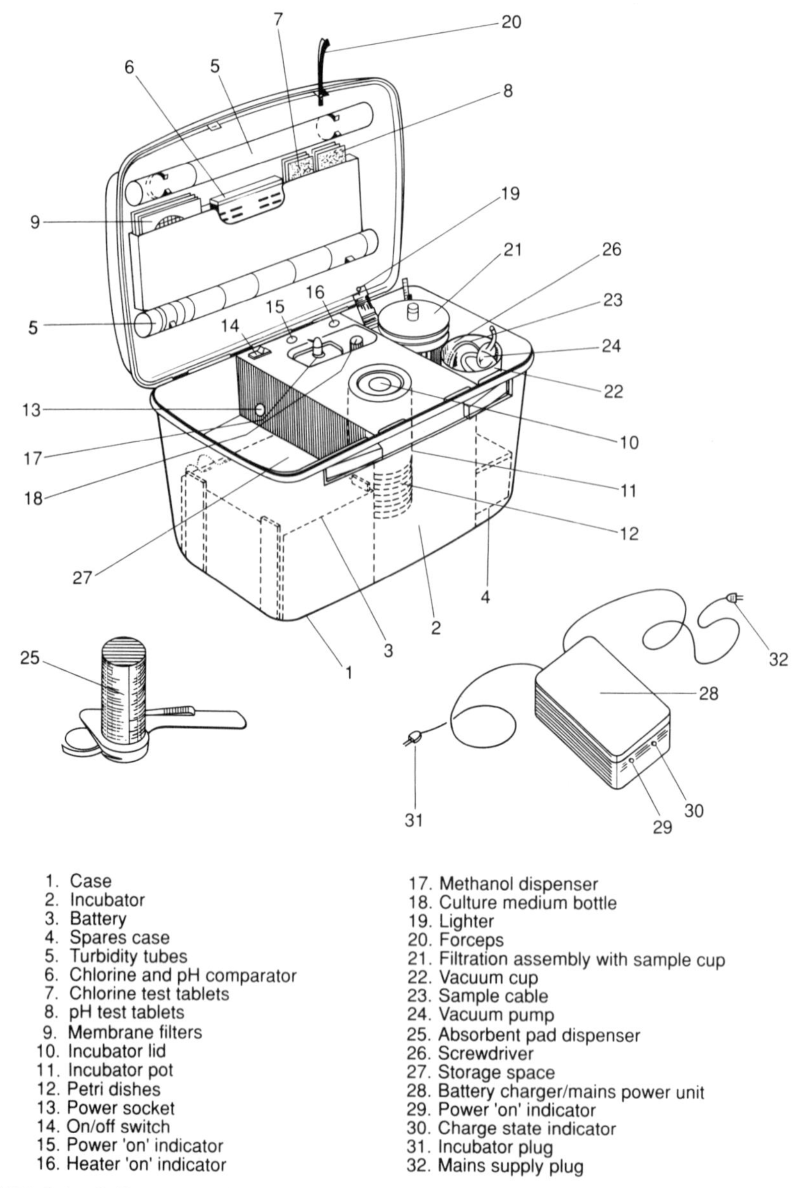

FIG. 6.1 Self-contained field portable water test kit for critical parameter testing (DelAgua Ltd, 1988).

groundwater protection is poor, extremely high levels of total coliform contamination may be demonstrated. It is usual to find >85 per cent of all rural supplies and often 100 per cent of dug wells contaminated with highly variable numbers of total coliform bacteria. Few technical staff are able to interpret the significance of these widely varying numbers of total coliform bacteria, and in many countries the complete confirmed coliform test is not done. For unchlorinated supplies, which constitute the majority of drinking water supplies in developing countries, the total coliform test is therefore of little discriminatory value. By contrast the faecal coliform test is readily interpreted in human risk terms.

If it is possible to take a source sample before chlorination, to assess bacterial contamination, this should be done. However, where sources are chlorinated the sanitarian or sampler should first take a water sample for the determination of chlorine residual on site (WHO 1988b).

The free residual chlorine should normally be greater than 0.2 mg/l and less than 1 mg/l, but this will depend on the distance from source to distribution system and on ambient temperature. The WHO guideline value for turbidity is <5 turbidity units (TU). If these two criteria are met it is highly unlikely that the sample will contain faecal coliform bacteria and therefore faecal coliform analysis becomes a precautionary procedure. Where results are not within the above criteria it is necessary to carry out faecal coliform analysis (WHO 1986).

The following methods and procedures of on-site critical parameter analysis are recommended as reliable, and were used in two of the three pilot projects. The field portable water test kit shown in Fig. 6.1 was used in Peru and Indonesia.

6.2 On-site physico-chemical measurements

6.2.1 Chlorine residual and pH analysis

1. Wash the comparator cells three times with the water which is to be analysed and finally completely fill all (three) cells with the sample to be tested.

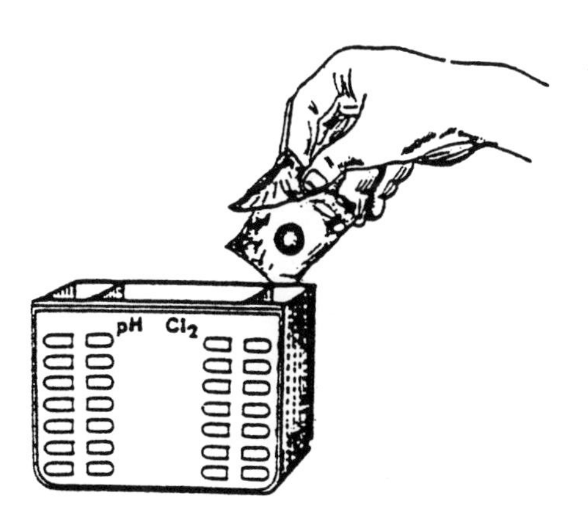

2. Drop a DPD (Diethylpara-phenylene diamine) No. 1 tablet into the right-hand cell, indicated Cl_2, and phenol red tablet in the left-hand pH cell.

3. Replace the lid of the comparator and push down firmly to seal. Invert the comparator repeatedly until the two tablets have dissolved completely.

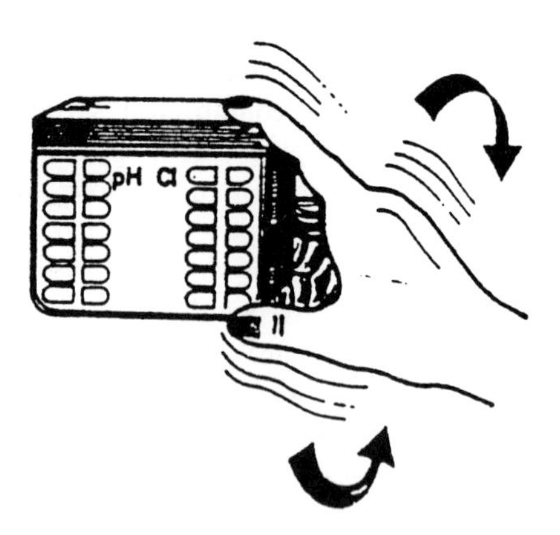

4. Immediately read the free chlorine residual and pH concentrations by holding the comparator up to daylight, and matching the colour developed in the cells with the standard colour scale in the central part of the comparator. If the colour falls between two standard colours then it will be necessary to estimate the concentration. Record the result on a report sheet in mg/litre.
5. To test for *total chlorine residual*, do not discard the liquid in the comparator, but remove the lid and add a tablet of DPD No. 3 to the same right-hand Cl_2 cell containing dissolved DPD No. 1 and replace the lid again.
6. Invert the comparator repeatedly to dissolve the tablet. This will take longer than the first. The colour will gradually intensify if combined chlorine is present, therefore read the colour developed after 10 minutes; this represents total chlorine residual in mg/litre.
7. Subtract the free chlorine result from the total chlorine result to obtain the combined chlorine concentration, thus:

DPD No. 1	= Free chlorine residual
DPD No. 1 + DPD No. 3	= Total chlorine residual
Total – free	= Combined chlorine residual

The combined chlorine may also be considered as the chlorine demand, that is the sum total of oxidizable inorganic (mainly Fe and Mn in groundwater and lesser amounts of inorganic nitrogen) and organic substances which have reacted with chlorine and deviated from the principal objective of microbial disinfection.

6.2.2 Turbidity determination

The range of the two-tube method is <4–2000 TU; the tubes are graduated using a logarithmic scale with the most critical values inscribed. It is possible to make approximations of intermediate values by interpolation. Groundwater should almost always be less than 5 TU; this is essential for effective disinfection.

1. Push the upper tube (open at both ends) squarely into the lower tube taking care not to crack the watertight seal. Ensure that good illumination is available; strong daylight is adequate. Observe that the black circle at the base of the tube is in clear, strong contrast to the white background.

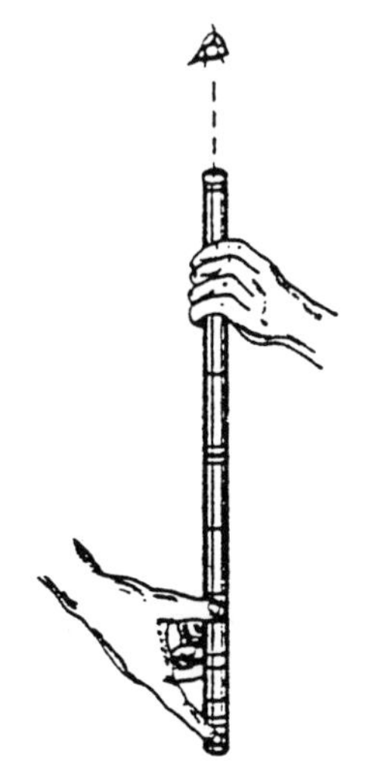

2. Pour the water sample into the tube from the sample cup, a little at a time, until the *black circle just disappears* when viewed from the top of the tube. To avoid biased results do not try too hard to see the black circle. Avoid making bubbles as these will produce artificially high readings. If it is impossible to avoid creating bubbles then wait until these have risen to the top of the tube before taking the reading.

6.3 Bacteriological field sampling

A sterile sample jug or bottle is required for bacteriological analysis.

6.3.1 Handpumps and boreholes

Handpumps are readily sampled from the pump mouth. *Boreholes* should be equipped with a *sampling tap* in the pump house and should be sampled as follows:

1. Remove any attachments from the tap, e.g. nozzles, anti-splash devices or pipes. Check that the tap does not leak and that all seals are in good condition.

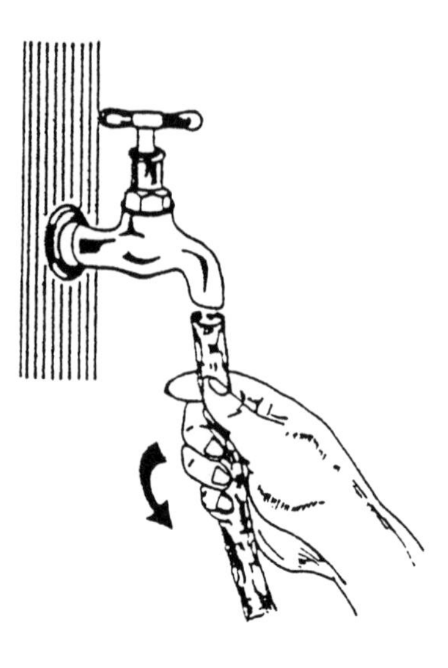

2. It is rarely necessary to flame a tap to remove gross contamination, but do check that the mouth of the handpump or sample tap is clean. Normally it is necessary to handpump or let the sample tap run to waste for at least 1 minute before taking the sample. This ensures that water unrepresentative of the main source is wasted out of the tap and associated pipe.

3. Collect the sample into a sterile sample cup or sterile bacteriological sample bottle. Since the standard faecal coliform count is expressed as a colony count per 100 ml, at least twice this amount should normally be collected.

6.3.2 Open dug wells

1. *Open wells* may be sampled by stainless steel sample cup. Use a stainless cable fastened to a hole in the lip of the cup by means of the clip on the end of the cable.

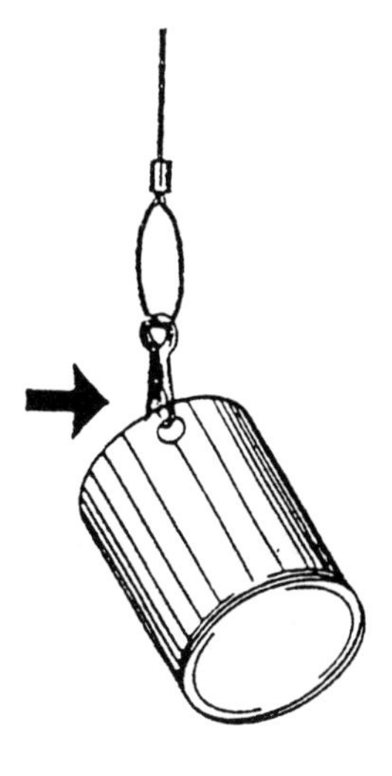

2. It may be necessary to increase the length of the cable by attaching a rope or string to the sample cable. *Take care* not to lose the sample cup by tying it securely.

3. Lower the cup into the well, avoiding contamination by taking care not to touch the walls of the well.

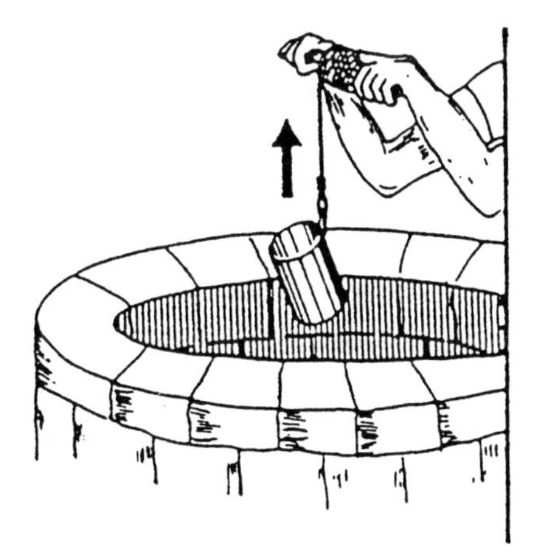

4. The sample may now be analysed on site, or transferred to the district laboratory in a sterile sample bottle in a cold box for analysis the same day.
5. If it is necessary to transfer samples in bottles, special bottle caps with protective sleeves should be used. These avoid the high risk of contamination transfer to the sample via the cap to the neck of the bottle.

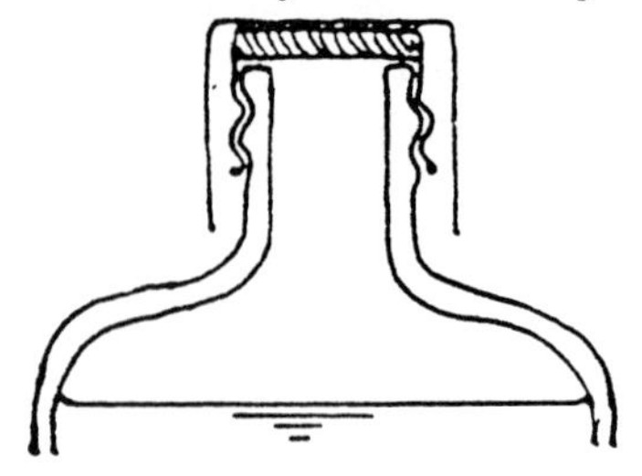

6.4 Faecal coliform analysis on site

The only means of carrying out on-site faecal coliform analysis is by membrane filtration; this is in any case more accurate than the multiple tube method and carefully executed provides results equivalent to those achievable in a central reference laboratory (DelAgua 1988).

6.4.1 Preparation of culture medium

An ideal culture medium is Oxoid's Lactose Sodium Lauryl Sulphate Broth which has a long shelf life. Dissolve 76.2 g of medium in 1 litre of distilled water, distribute in bottles and autoclave. Once a bottle of medium has been opened it should preferably be used the same day to avoid contamination. For this reason prepare small amounts for sterilization and storage for up to 6 months in 30–50 ml bottles.

1. Using an absorbent pad dispenser (or sterile forceps), place one pad into each Petri dish using aseptic technique. This may be done at the base or laboratory, before leaving for the sampling site, if preferred.

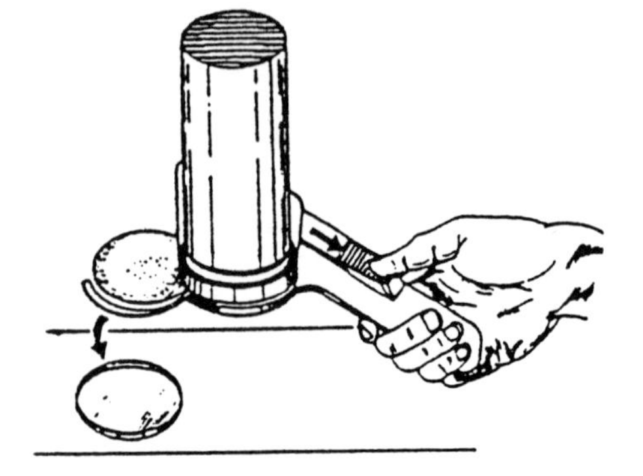

2. Pour enough faecal coliform culture medium on to the pad to soak it completely and provide a slight excess (2.5 ml). Replace the bottle cap immediately. Do not allow the bottle neck to come into contact with any external objects. This may also be done before going to the field. Replace the lid on the Petri dish.

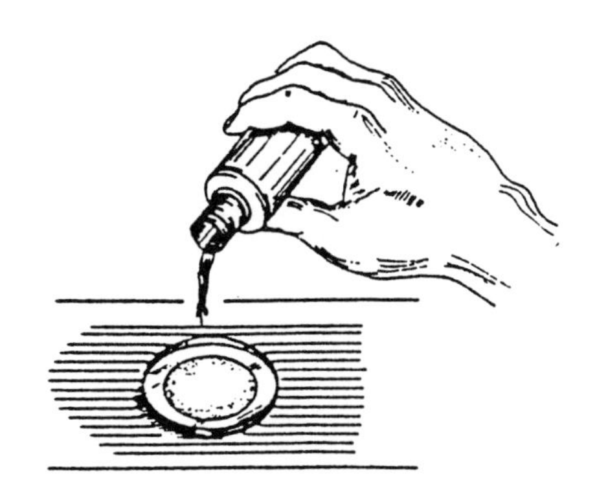

3. Immediately before processing a sample, drain off most of the excess medium by inverting each dish with the lid removed. Always leave a slight excess to prevent drying out of the pad during incubation.

6.4.2 Preparing for sample filtration

4. The DelAgua test kit is provided with all the equipment necessary for faecal coliform testing in the field. This includes field sterilizable sample cup and filtration assembly which are resterilized between each sample and therefore ready for use on arrival at each new sample site.

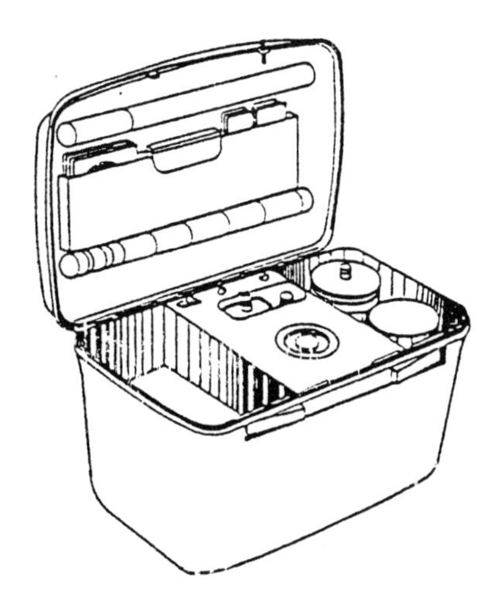

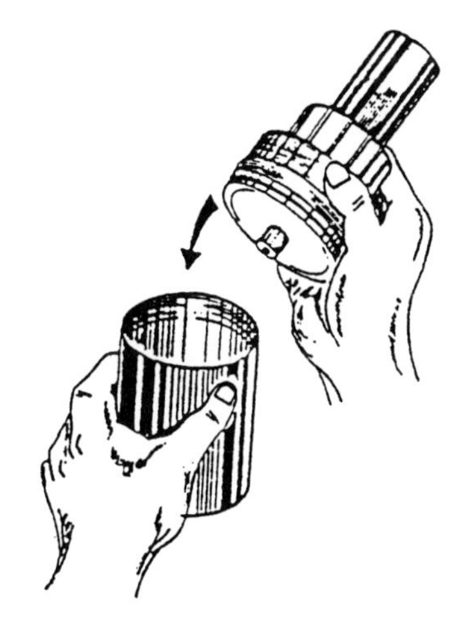

5. The sterile filtration assembly travels inverted into the sample cup. At arrival on the sampling site push the filtration assembly base on to the vacuum cup and place it in an upright position in the kit where it is convenient for filtration. Do not place the apparatus on the ground where it may be contaminated with soil.

6. Only remove the sample cup when ready to take the sample. This reveals the top of the sterile filter funnel. Unscrew the filter funnel collar to check that it is easy to lift out and ready to receive a sterile membrane filter, but do not place the filter funnel on any surface other than the filter base.
7. Flame the tips of the forceps with a gas lighter and allow to cool.

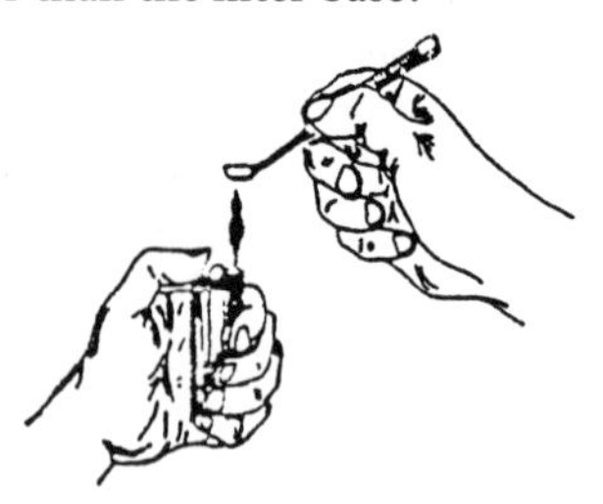

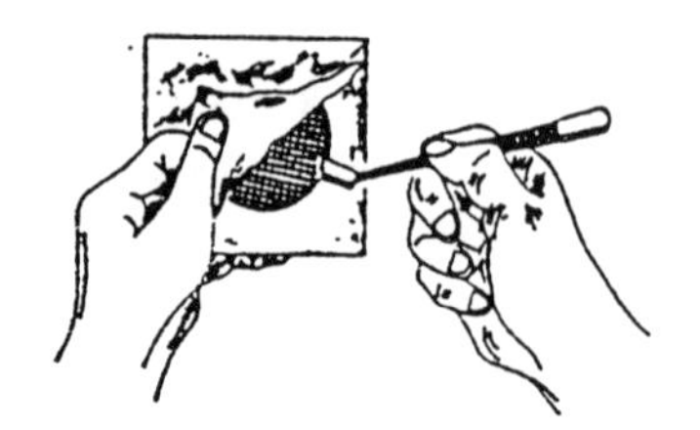

8. Remove a sterile membrane from its packet by the edge, using the sterile forceps.

9. After lifting the filter funnel off the base the membrane may be transferred on to the bronze disc filter support. Replace the funnel and collar immediately without allowing them to come into contact with any external objects.

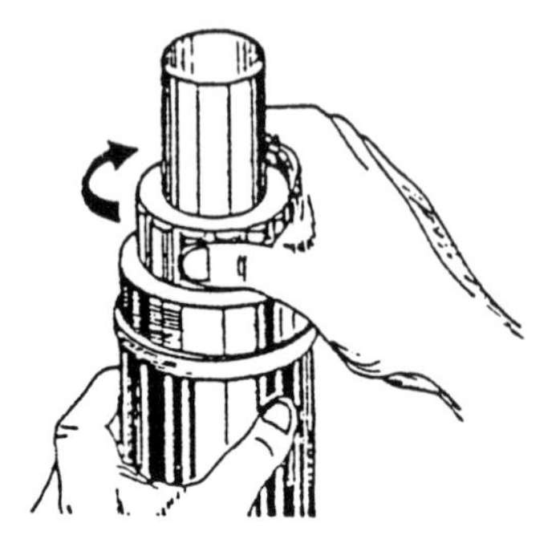

10. Screw the collar down tight to hold the membrane in place and provide a watertight seal.

11. Pour the sample from the sample cup into the appropriate mark engraved on the internal surface of the funnel (usually 100 ml for groundwater samples).

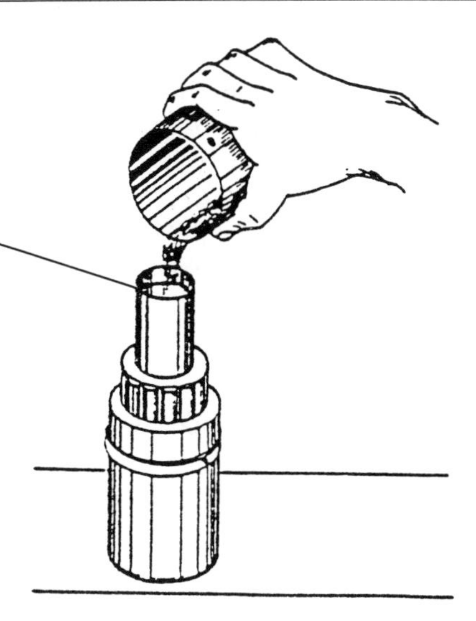

6.4.3 Sample processing and resterilization of apparatus

12. Insert the vacuum pump connector into the vacuum connection in the filtration base. Squeeze the pump to draw a sufficient vacuum to suck all the water through the filter and then disconnect the pump.

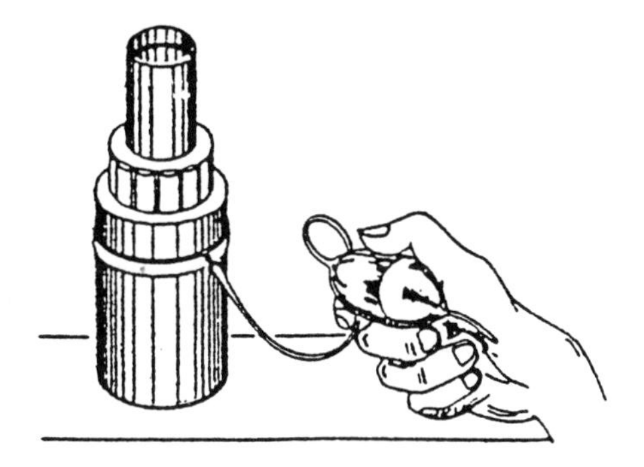

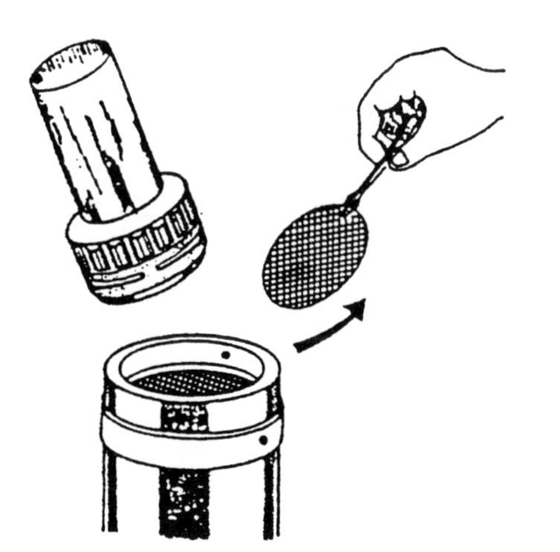

13. Unscrew the collar and remove the funnel and collar with one hand. Using the sterile forceps lift the membrane by the edge from the filtration base.

14. Transfer the membrane, grid side uppermost, to a Petri dish containing an absorbent pad soaked in culture medium. Lower the membrane gently on to the pad to avoid trapping air bubbles under the membrane.

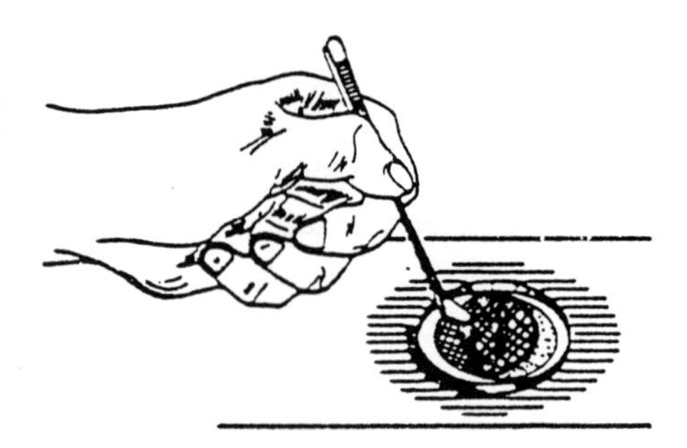

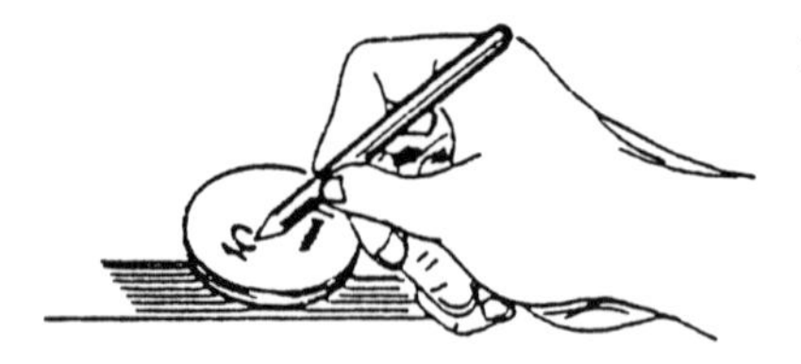

15. Replace the lid of the Petri dish and mark the dish with sampling data (volume filtered, source and sample number) using a wax or soft lead pencil.

16. Place the Petri dish, with the lid uppermost, into the carrier and return the carrier to the incubator pot. Replace the incubator lid.

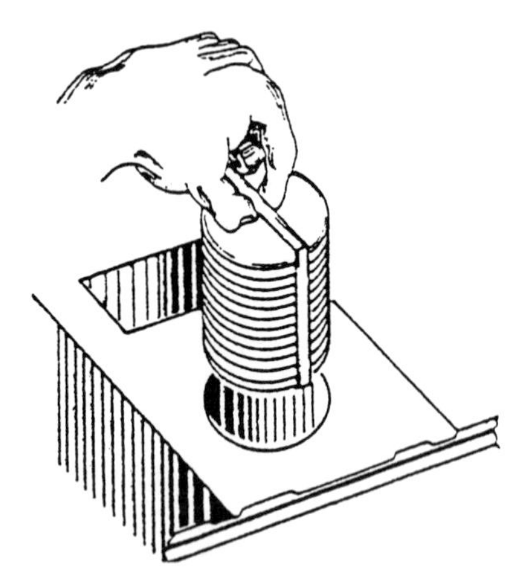

17. In order to resterilize the sample cup, the residual sample must first be discarded and the cup and filtration assembly dried with a clean, dry towel or tissue.

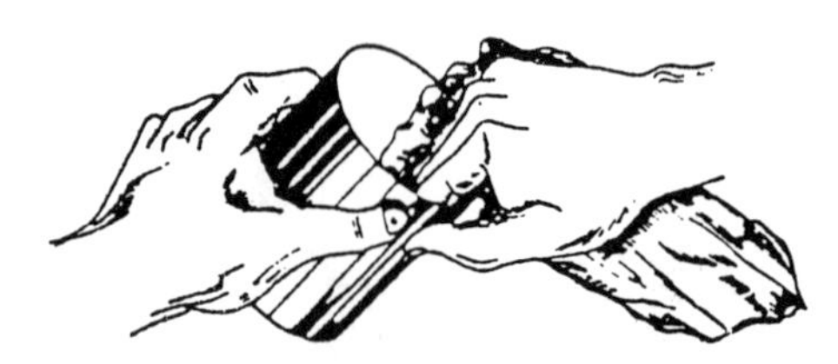

18. Reassemble the filtration assembly loosely and squirt approximately 1 ml of methanol into the sample cup.

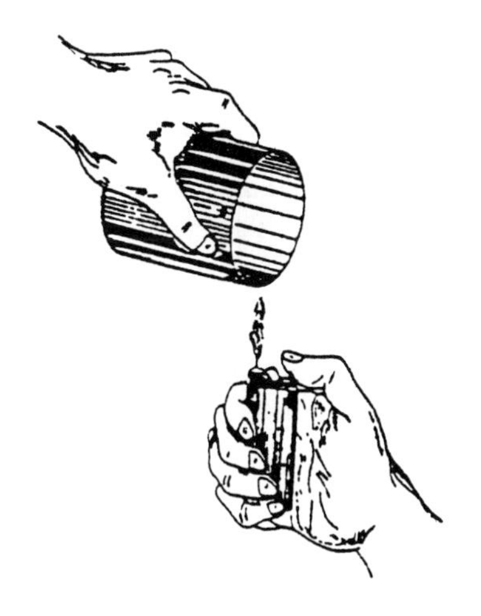

19. Ignite the methanol in the sample cup using the gas lighter. Avoid burns by keeping the mouth of the cup away from the operator and bystanders. Allow the methanol to burn in the cup on a flat surface in the kit. After several seconds, invert the filtration assembly firmly into the sample cup *before* the methanol has burnt out.

20. Keep the filtration assembly and sample cup closed in transit so that it is sterile and ready for use on arrival at the next sample site.

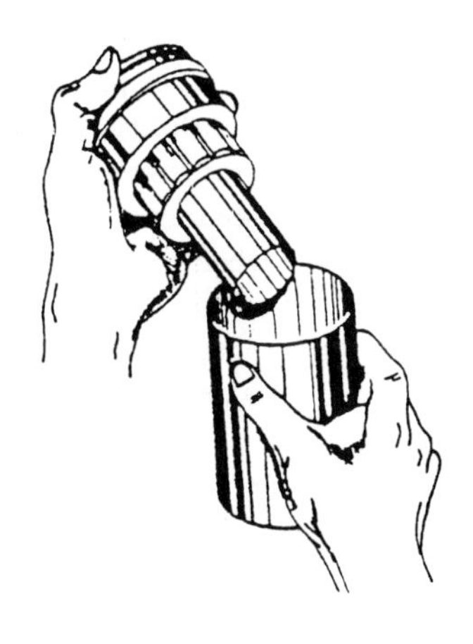

6.4.4 Incubation of samples and enumeration of faecal coliforms

Once the last sample of the day has been taken and placed in the incubator it may be switched on immediately in a hot climate. In a cold or temperate climate switching on the incubator should be delayed for between 0.5 and 1 hour to avoid temperature shock. The samples are incubated for 14–18 hours at a temperature of 44 °C. Instructions for care of the incubator are provided in the DelAgua manual (DelAgua 1988).

Remove the Petri dishes after incubation and examine them in good incident light. All yellow colonies which have grown, typically having a diameter of 1–3 mm, are counted as faecal coliform bacteria on the medium previously specified. They produce acid from lactose on media in which the indicator is phenol red. Do not count colonies which are transparent or pink/red on cooling: these are non-lactose fermenters and are not coliforms.

Large numbers of yellow colonies can be accurately counted by methodically moving a pointer above the grid lines. It is difficult but possible to count more than 200 colonies on a membrane; where the approximate level of pollution cannot be estimated it may be necessary to use several sample volumes and then it is necessary to convert the number of colonies counted to count/100 ml as follows:

Volume filtered	*Faecal coliform count/100 ml*
100 ml	No. colonies/membrane × 1
50 ml	No. colonies/membrane × 2
10 ml	No. colonies/membrane × 10

The significance of faecal coliform counts is discussed in section 7.1.1.

6.5 Tracing sources of pollution

Unfortunately in many countries the problems of pollution are so obvious, widespread and overwhelming that it is more often a question of which of the various sources of pollution should be plugged first and for how many systems is there sufficient budget to make a worthwhile improvement. It has already been demonstrated in Chapter 5 that, at least for periodic sanitary surveys, it is relatively straightforward to draw up a finite list of potential points of pollution of the source for investigation, and that the potential sources of pollution are relatively easily identified. The methodology is now therefore at the stage of application when detailed lists of polluted sources can be prepared for evaluation as described in step 7 (Fig. 4.3). However, sophisticated methods may sometimes be necessary to identify a source (or trace a route) of pollution before remedial action is undertaken.

Occasionally, it is essential to confirm the route of movement of contaminants through water in order to assess how and from where sources are being contaminated. Although fluorescent dyes and lithium have sometimes been used to do this, in recent years bacteriophages (bacterial viruses which have no health effect on animals) have been used for this purpose (Wimpenny *et al.* 1972). Bacteriophages such as those of *Serratia marcescens*, which are uncommon in the natural environment, are particularly suited to the tracing of water flow paths since they are relatively cheap and can be readily cultivated to high concentrations (Skilton and Wheeler 1988).

CHAPTER 7

Evaluation of Surveillance Results

7.1 Classification of results

The difference between the outcome of water analysis and the sanitary survey is that analysis will detect actual contamination and the level of contamination, whereas the sanitary survey which includes analysis should identify the points of the system at risk. The two activities are complementary and interdependent and together they should confirm that pollution is occurring. Together they should also, in most cases, identify the source of pollution. Sometimes, however, more sophisticated techniques are required to identify the source of pollution.

7.1.1 Proposed bacteriological grading schemes

It has already been pointed out that the majority of rural supplies are unchlorinated and it is therefore inevitable that they will contain large numbers of total coliform bacteria which may have limited faecal significance. It is therefore recommended that the bacteriological classification scheme be based primarily on the thermotolerant faecal coliform bacteria.

In order to distinguish between water sources and systems which conform to WHO guidelines for faecal contamination (zero *Escherichia coli*/100 ml) and those with different levels of contamination, a robust classification system is proposed based on increasing orders of magnitude of faecal coliform contamination as given in Table 7.1.

Table 7.1 Indonesian *E. coli*/faecal coliform classification scheme for water supplies

Grade	Count/100 ml	Risk
A	0	No risk
B	1–10	Low risk
C	11–100	Intermediate to high risk
D	101–1000	Gross pollution; high risk
E	>1000	Gross pollution; very high risk

Table 7.2 Peruvian *E. coli*/faecal coliform classification scheme for water supplies

Grade	Count/100 ml	Risk
A	0	WHO Guideline value; no risk
B	1–10	Low risk
C	11–50	Intermediate to high risk
D	>50	Gross pollution, high risk

The scheme presented in Table 7.1 accommodates the high levels of contamination found in Indonesian point source rural groundwater supplies: it is a modification of the scheme used in the Peruvian pilot project shown in Table 7.2, and published by Lloyd (1982). The A–D classification scheme as shown in Table 7.2 was applied to Peruvian piped supplies and was adequate because few supplies delivered water with >50 faecal coliforms. It is therefore recommended that surveillance programmes should begin with the Indonesian scheme and transfer to the Peruvian scheme when remedial action programmes have succeeded in improving the grade E supplies.

7.1.2 Proposed sanitary survey grading

The sanitary survey forms present a number of technical points requiring consideration in using the risk information to improve effectively the organization and safety of water supplies including the following:

1. How can the data be converted into relative risk to compare a collection of systems and subsequently into simple remedial measures at local level?
2. How many 'false positives', i.e. mistaken risk points, can be tolerated without invalidating the scoring system? In other words is the scoring system robust?
3. How can a scoring system be developed which is sufficiently discriminatory to separate out systems requiring urgent attention, without overwhelming the workforce with the sheer size of the remedial action? (There is, for example, little advantage in a strategy which classifies 80 per cent of systems as being at 'very high risk' unless massive resources are available for remedial action.)
4. How can the most important source(s) of pollution be identified when a number of potential sources may have been identified?

In order to tackle these problems it is necessary to use all the available information systematically. That which is available for each source examined includes basic analytical data, mainly faecal coliform counts together with corresponding sanitary risk scores.

It has already been pointed out in Chapter 5 that for each type of water source the number of risk points recorded during the sanitary inspection may be totalled to give a sanitary risk score. The unweighted totals of sanitary inspection risk points (1–>10) have been used to produce a risk score for each installation. These scores are arbitrarily graded into different levels of relative risk as indicated in Table 7.3.

Table 7.3 Sanitary inspection risk scores

Risk score	Risk
0	No risk
1–3	Low risk
4–6	Intermediate to high risk
7–>10	Very high risk

It will be appreciated that it is a relatively simple matter to grade point source systems where there are only 10–13 points for inspection. It is more complicated to grade larger community water supply systems which sometimes include a number of sources, treatment plants, reservoirs and a distribution system. For this reason the results of the three pilot projects are presented separately, particularly since the Peruvian supplies were exclusively piped systems.

7.1.3 Combined evaluation approach

By examining the faecal grading together with the sanitary inspection risk scores for a large number of facilities it should be possible to assess whether water quality and the risks identified by inspection are broadly correlated. This may be done in monthly reports at the district surveillance office for local remedial action, or less frequently by regional or national offices for strategic planning purposes. It is then necessary to test and inspect various groupings and gradings shown in Fig. 7.1 to assess whether they make sense. The faecal grading (A–E) has been combined with the sanitary inspection risk grade to produce zones for recommending remedial action in Table 7.4.

With this simple zoning system as presented in Fig. 7.1 there is an underlying assumption that increasing sanitary risk will tend to be combined with increasing levels of faecal pollution. If this general assumption is correct a plot of many individual systems of different types will produce a broad band of points from top right (for systems requiring urgent action) to bottom left (for systems requiring no action) shown theoretically in Fig. 7.2.

It is implicit in this assumption that few facilities will be found to occur in the top left or bottom right of the graph. For the few sources that do occur in these areas, the shape of the zoning in the graphs allows a reciprocal arrangement with respect to combined risk factors assessment and subsequent action. On the one hand a high sanitary risk score, of say 8, occurring with low-level faecal contamination, say grade B, still requires urgent action. On the other hand, a

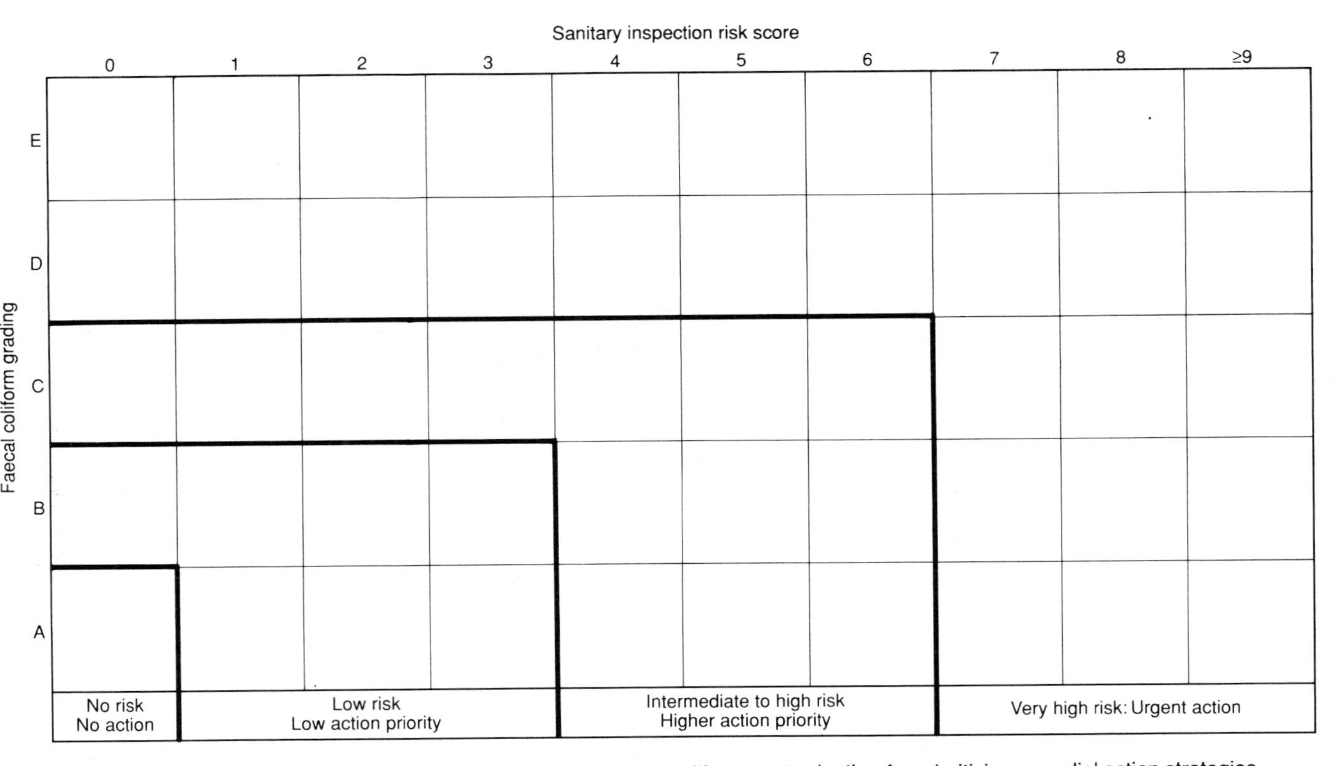

FIG. 7.1 Zoning of combined risk analysis of inspection and faecal coliform contamination for prioritizing remedial action strategies.

Table 7.4 Combined risk analysis of sanitary inspection and faecal coliform contamination

Faecal grade	+	Sanitary risk grade	=	Action priority
A	+	No risk	=	No action
B	+	Low risk	=	Low priority
C	+	Intermediate to high risk	=	Higher priority; as soon as resources permit
D/E	+	Very high risk	=	Highest priority; most urgent action

low sanitary risk score, of say 3, occurring with grade D faecal contamination, also requires urgent action. In fact the hypothesis of top right to bottom left holds good so far for 90 per cent of 500 water sources studied. The 10 per cent of odd results outside the broad band are discussed later.

The theoretical distribution shaded in Fig. 7.2(a) is an *ideal* correlation between sanitary risk and faecal grade in which the two assessments are always strongly correlated. In reality it has been found that the data are distributed more broadly as shown in Fig. 7.2(b). This is still a very useful correlation because it demonstrates that the sanitary inspection is robust (compared with bacteriological analysis). In this model sanitary risks are frequently identified even in the absence of faecal contamination (grade A). This is particularly important when the surveillance staff are dependent on a single bacteriological analysis. *It is worth emphasizing that the analysis is representative of one moment in time, whereas the inspection takes account of the entire previous history of the installation as well as future points of risk.* It would be most unsatisfactory if sanitary inspection routinely underestimated the risk of bacteriological and microbiological contamination.

By contrast, the theoretical distribution presented in Fig. 7.2(c) is considered to be *not* at all useful because faecal contamination is here occurring either in the absence of observable points of risk or when moderately contaminated with relatively many points of risk identified. Perhaps the most important point then is that the sanitary inspection should not only reveal actual points of pollution, but also predict potential incipient points of risk which can then be prevented from becoming actual points of pollution.

7.2 Indonesian results

When new provincial laboratories are established it is essential that their results be validated initially and then periodically by a competent reference laboratory. Using the revised scheme for both total and faecal coliforms the Indonesian pilot project laboratory carried out a laboratory comparison with the same samples being analysed by the National Reference Laboratory in Yogyakarta. The comparison presented in Table 7.5 makes use of a frequency matrix to compare

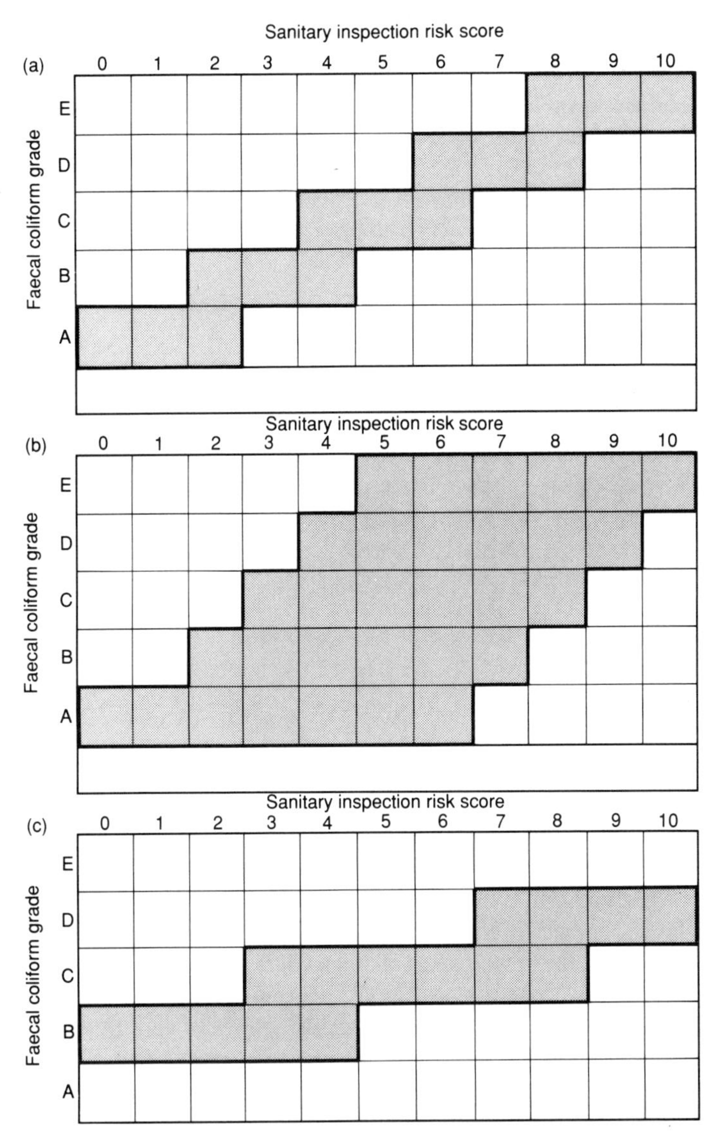

FIG. 7.2 Theoretical association between sanitary risk scores and faecal grading of rural water supplies: (a) ideal; (b) useful; (c) *not* useful.

the agreement between grades. It may be observed that (35 + 48) 83 per cent of all total and faecal coliform results fall within the top two levels of homogeneity indicating a reasonable level of agreement. It may be noted that interlaboratory comparisons in the USA are only marginally better than this. Equally importantly, when the simpler A–D classification was applied (not shown here), 88 per cent fell within the top two levels of homogeneity. Thus although the A–E scheme loses only 5 per cent in discriminatory value compared with the simpler

A–D classification, the A–E scheme none the less permits the separate grouping of the most grossly contaminated water supply facilities which logically should receive the most urgent attention for remedial action, and so the more extensive scheme was recommended for Indonesia.

Table 7.5 Total/faecal coliform A–E classification comparison of all replicate samples between provincial Hospital Lab. and National Reference Lab. Yogyakarta, Java

Homogeneity level						Total at each level No. (%)
1	AA	BB	CC	DD	EE	
	3	2	9	4	10	28 (35)
2	AB	BC	CD	DE		
	9	5	13	11		38 (48)
3	AC	BD	CE			
	3	3	5			11 (14)
4	AD	BE				
	1	1				2 (3)
5	AE					
	0					0 (0)
No. of cross-checks	16 +	11 +	27 +	15 +	10 =	79 (100)

In the first year of the Indonesian project Wardojo (1987) reported that the analytical results from 2546 samples demonstrated that 80 per cent of facilities were classified as faecally contaminated but no complementary sanitary inspection was done. In the second year, of 500 facilities examined, 77 per cent were classified as faecally contaminated and 72 per cent of facilities were classified as bad on subjective sanitary inspection criteria. In the third year, of 1000 facilities examined, 66 per cent were faecally contaminated and 67 per cent had a bad sanitary inspection result. Unfortunately, the association between faecal contamination and sanitary inspection could not be correlated statistically mainly because the faecal grading and sanitary risk scoring schemes had not been introduced. It was therefore not until the final phase of the pilot project that the inspection report forms were fully revised, translated into Indonesian and used as described below. This provided a more accurate picture of the level of contamination and sanitary status of the main types of point source water supplies.

Table 7.6 summarizes the faecal grading of 328 point source facilities investigated in 1988 and highlights the superior quality of protected groundwater. It demonstrates that tubewell water is least contaminated, 86 per cent of shallow tubewells and 84 per cent of deep wells falling in categories A + B. At the other end of the scale 9 per cent of these deep wells and 7 per cent of shallow wells fall in the very high risk grades (D + E). This contrasts with 22 per cent open dug wells and 20 per cent of converted dug wells in the very high risk grades, while only 12 per cent of rainwater tanks were contaminated at these levels.

Table 7.6 Faecal coliform contamination of point source drinking water facilities in the pilot project area in Java

		Grade of faecal coliform contamination*				
Type of facility	Total	A (%)	B (%)	C (%)	D (%)	E (%)
1. Unimproved dug well	55	3 (5)	32 (54)	7 (12)	7 (12)	6 (10)
					(22) (D+E)	
2. Converted handpumped dug well	97	20 (20)	31 (32)	26 (27)	19 (19)	1 (1)
					(20) (D+E)	
3. Handpumped *shallow* tubewell	100	72 (72)	14 (14)	7 (7)	2 (2)	5 (5)
		(86) (A+B)				
4. Handpumped *deep* tubewell	44	26 (59)	11 (25)	3 (7)	3 (7)	1 (2)
		(84) (A+B)				
5. Rainwater tanks	32	2 (6)	10 (31)	16 (50)	3 (9)	1 (3)
					High risk Gross pollution MOST URGENT ACTION	
Total facilities under study	328					

* Grade *E. coli* faecal coliform count/100 ml:
A = 0 (WHO Guideline recommendation); no risk
B = 1–10; low risk
C = 11–100; intermediate to high risk
D = 101–1000; gross pollution; high risk
E = >1000; gross pollution; very high risk

All the matched bacteriological and sanitary inspection data have been combined and summarized in Table 7.7. The most obvious conclusion from this is that the facility presenting the highest risks are the 88 per cent of unimproved open dug wells, made up of the 43 per cent intermediate to high risk and 45 per cent very high risk categories. What is worrying is that the expense of conversion of dug wells by fitting a handpump and a 'sanitary' cover does not make a more substantial reduction in risk; 83 per cent are still in the two high risk categories.

Table 7.7 Combined risks for sanitary inspection and bacteriological analysis of 244 point source drinking water facilities in Gunung Kidul, Java

Type of facility	No risk No. (%)	Low risk No. (%)	Intermediate to high risk No. (%)	Very high risk No. (%)
1. Unimproved open dug well	2 (6)	2 (6)	15 (43)	16 (45)
			(88)	
2. Converted handpumped dug well	1 (1)	15 (16)	50 (53)	28 (30)
			(83)	
3. Handpumped *shallow* tubewell	0 (0)	2 (5)	30 (80)	6 (15)
4. Handpumped *deep* tubewell	5 (11)	20 (62)	8 (18)	4 (9)
5. Rainwater tanks	0 (0)	4 (12)	16 (50)	12 (38)
Remedial action	No action	Low action priority	Higher action priority	MOST URGENT priority

The similar high risk factors of converted, handpumped dug wells and unimproved, open dug wells shown in Table 7.7 are also reflected graphically by the similarity between the cluster distribution of facilities plotted in Figs 7.3 and 7.4. It was clear from supervisory site visits that little attention was paid to rendering the lining of the well when the conversion investment to handpumped dug well was made and consequently many of the same sources of pollution which affected unimproved open dug wells persist. It is vital that the sanitarian should raise the well cover and check the lining walls when the inspection is made.

Figures 7.3–7.7 demonstrate how the faecal coliform grades A–E and the sanitary inspection risk scores can be combined, presented graphically to show the association between observed risk and measured faecal contamination for each type of facility, and zoned for remedial action. The figures show at a glance how the different facilities cluster characteristically. The deep tubewells are typically well protected from sanitary risks and thus low sanitary risk scores predominate. The risk scores correlate well with a high proportion of A + B category bacteriological analysis, leading to clustering of results in the bottom left-hand corner of the graph (Fig. 7.6). By contrast, the converted dug wells (Fig. 7.4) produce a dense cluster in the intermediate to high risk zone and a broad band correlation from top right to centre, but almost no facilities with no risk in the bottom left of the graph. From a practical point of view the figures are more useful than the summary Table 7.7 because the figures include the facilities' code and therefore the sources requiring most urgent action can be immediately identified.

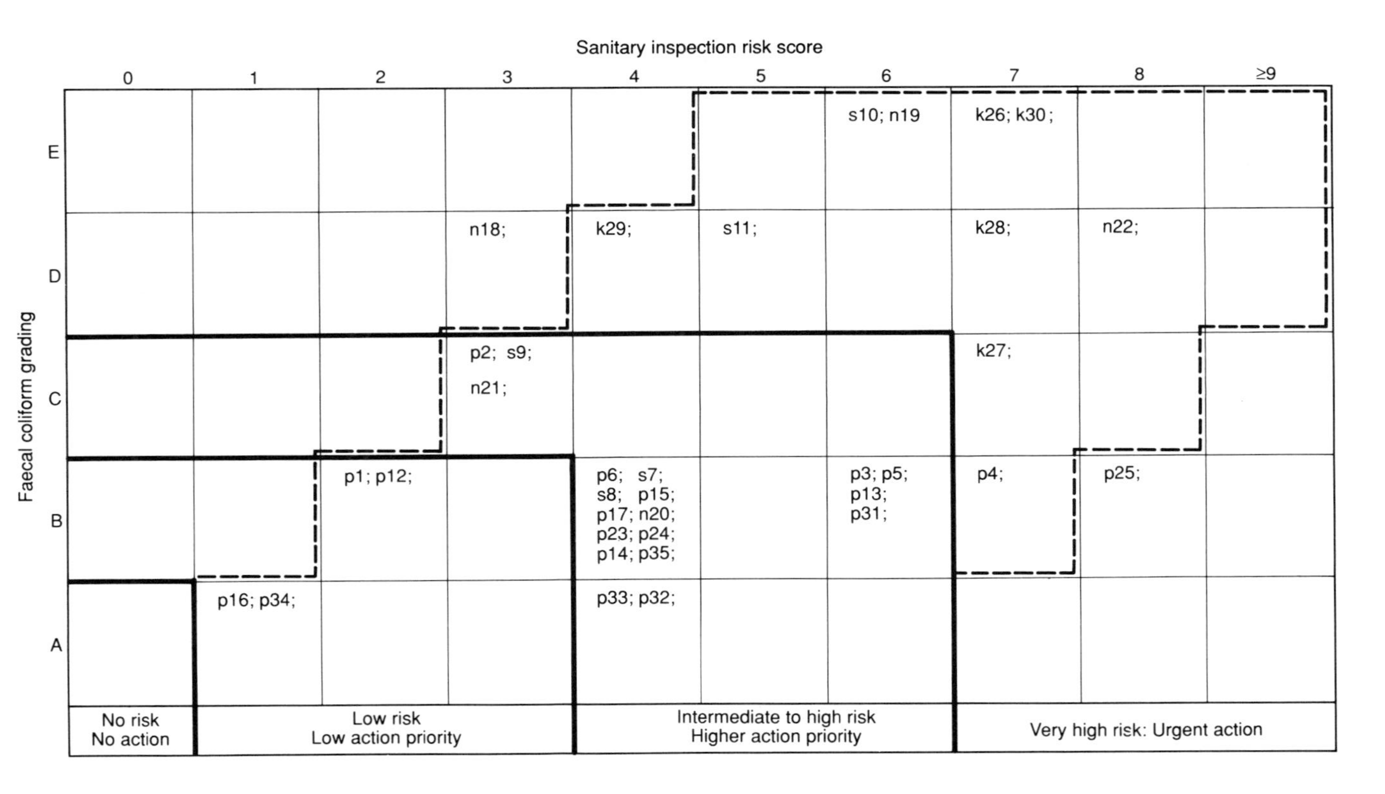

FIG. 7.3 Combined risk analysis of sanitary inspection and faecal coliform contamination of a single point drinking water facility – unimproved open dug wells. Survey period – June–August 1988; study area Gunung Kidul, Java. Code: letter = health centre; number = sample sequence.

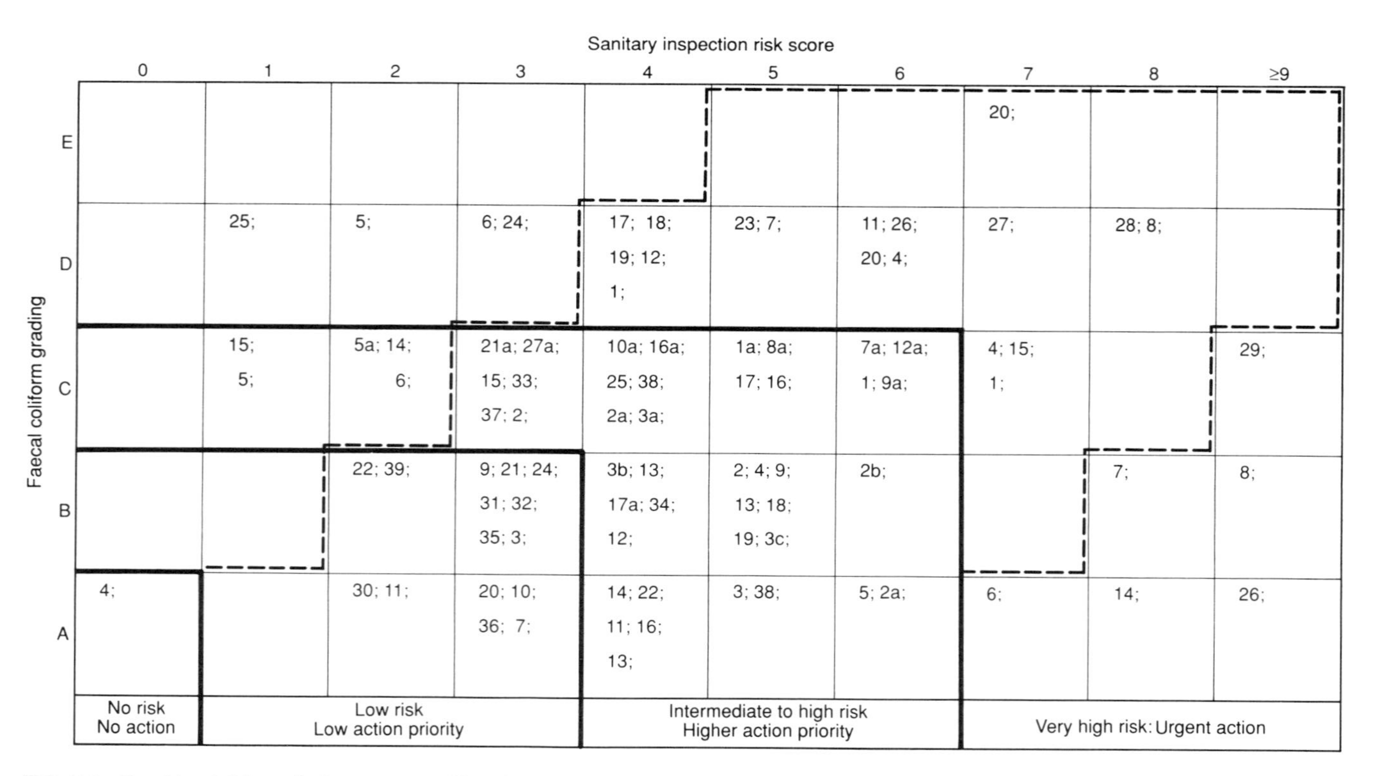

FIG. 7.4 Combined risk analysis – converted handpumped dug wells. Letter = location, other details as Fig. 7.3.

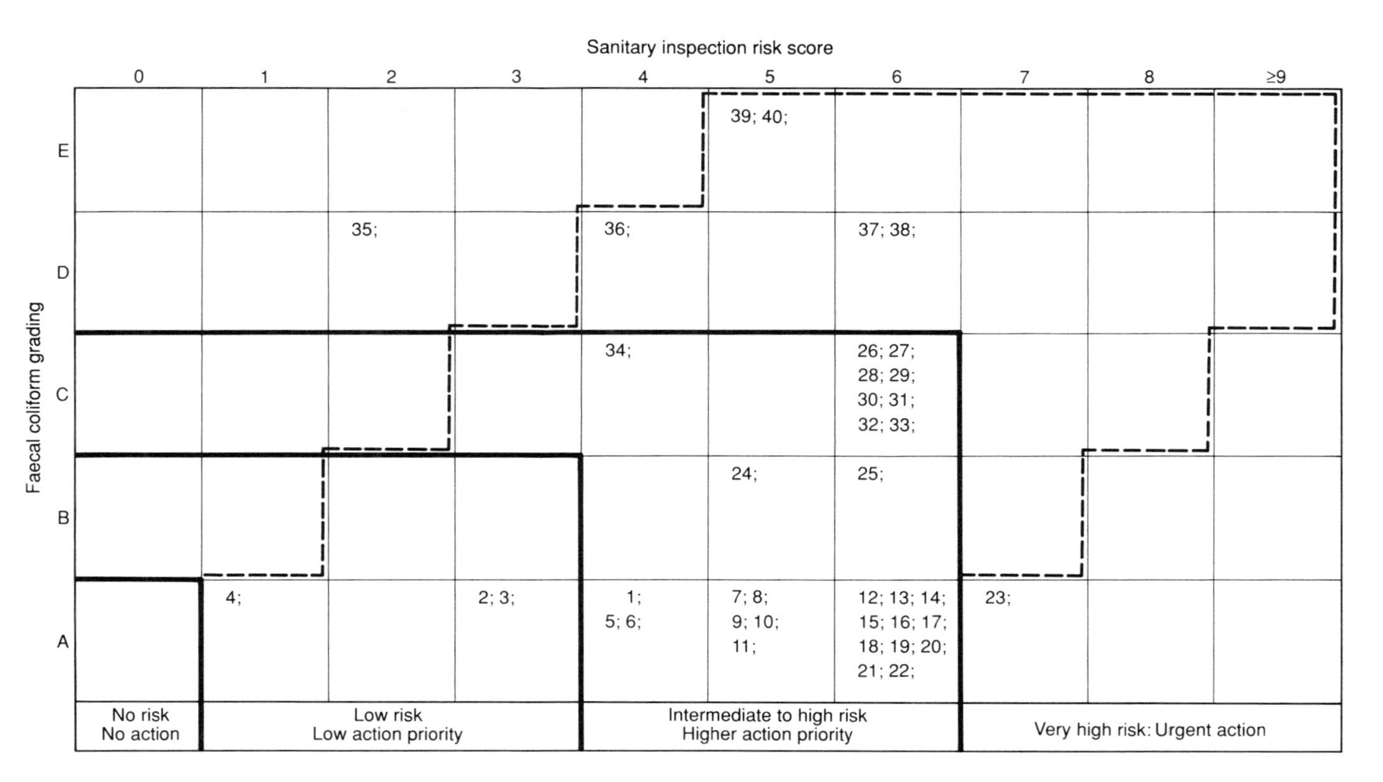

FIG. 7.5 Combined risk analysis – handpumped shallow (<10 m) tubewell. Number = sample sequence and location, other details as Fig. 7.3.

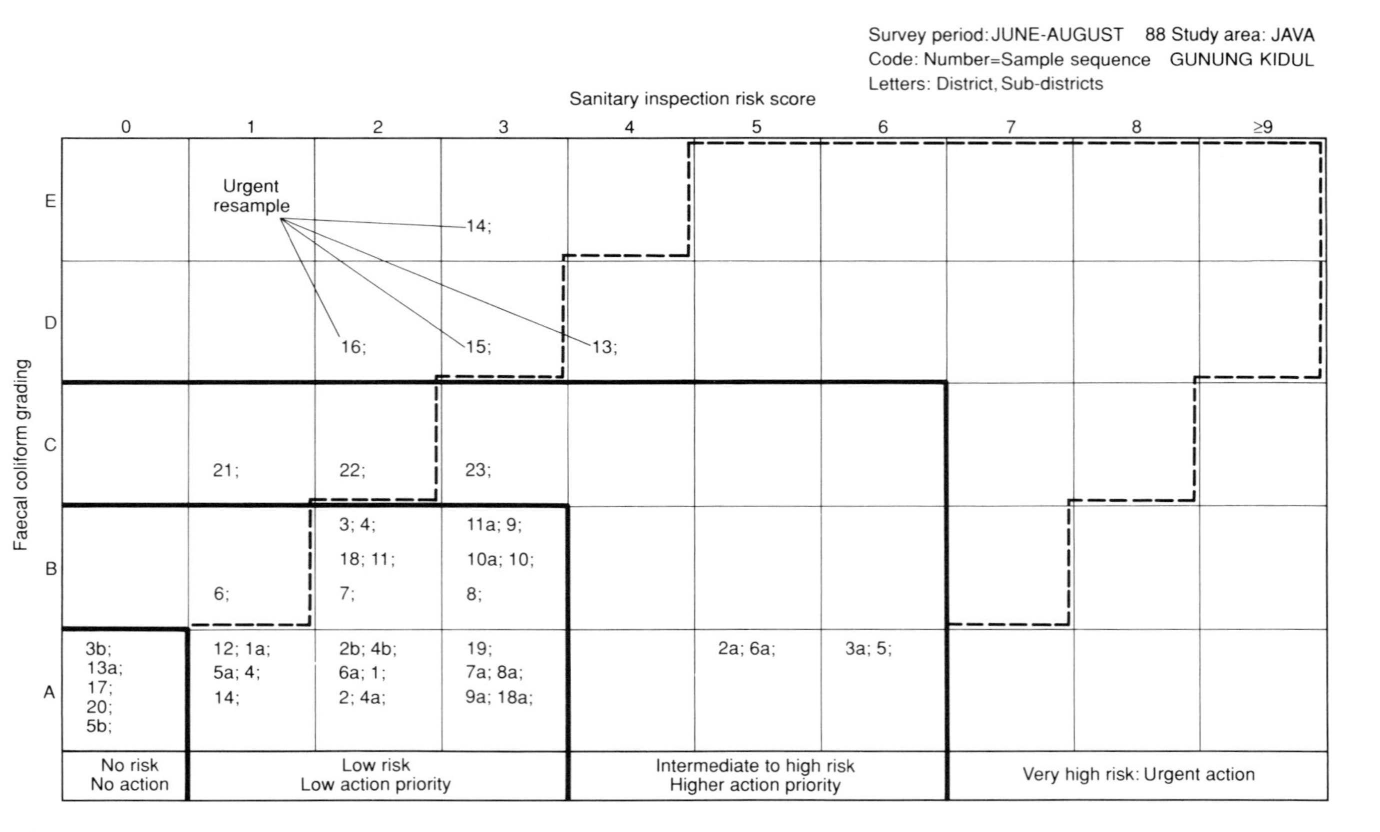

FIG. 7.6 Combined risk analysis – handpumped deep (>10 m) tubewell. Number = sample sequence, other details as Fig. 7.3.

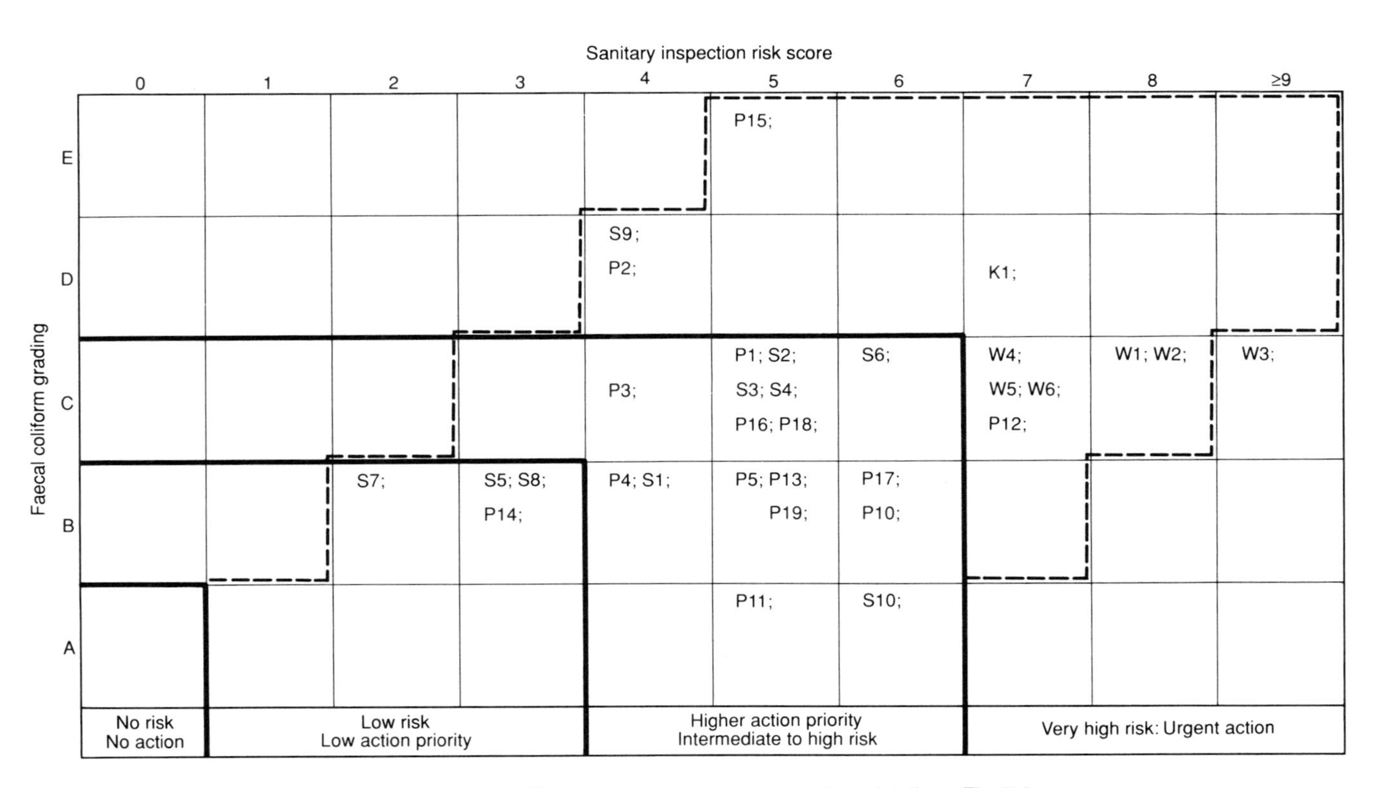

FIG. 7.7 Combined risk analysis – rainwater tanks. Number = sample sequence, other details as Fig. 7.3.

Compared with handpumped dug wells it appeared that a significant improvement (risk reduction), is achieved by shallow tubewells which are assessed at only 15 per cent in the very high risk category; however, the highest proportion (80 per cent) are classed in the intermediate to high risk category. When we examined bacteriological quality alone (Table 7.6) it was observed that 72 per cent were in grade A (WHO guideline value) and a further 14 per cent were in grade B, making a total of 86 per cent low risk with respect to faecal contamination alone; however, the clustering of points observed for the shallow tubewells in Fig. 7.5 indicates a poor correlation between level of faecal contamination and sanitary risks. Thus the largest cluster of A graded facilities are found in association with a high sanitary risk score of 6 (square A6). This may be due to the following reasons:

1. The facilities are deteriorating structurally with age, but have not yet reached the point where they are permitting contamination to enter the tubewell.
2. Several risk points in the sanitary inspection report form are being unnecessarily scored by the sanitarian.
3. Some of the risk points are not important sources of pollution in facilities as well protected as the tubewells.
4. The tubewells and dug wells may actually be installed in quite different pollution risk environments. For instance, if only dug wells are used in shallow water-table situations and boreholes where the water-table is deeper, for a given subsurface contaminant loading, there will be an intrinsically higher pollution risk in the dug well situation, because the unsaturated zone provides scope for dilution, retardation and elimination.

Whichever of these is the case, it serves to demonstrate that the sanitary reporting procedure is rigorous but requires follow-up. These apparent anomalies should therefore be verified by the district supervisor by rechecking the facilities in question. Whatever the outcome the results confirm the considerable bacteriological source protection in granular aquifers afforded by the drilled tubewell.

By contrast with the shallow tubewells, the combined risk analysis in Table 7.7 and clustering shown in Fig. 7.6, for deep tubewells, are very much closer to the situation which would be predicted from first principles. Only 9 per cent are very high risk and 18 per cent in the intermediate to high risk group, thus leaving 73 per cent in the low risk group. This correlates well with the 16 per cent (C + D + E) high risk and 84 per cent no risk and low risk bacteriological grades A (59 per cent) + B (25 per cent) and lends further strong support to the case for installing tubewells. The four points in Fig. 7.6 which are grossly contaminated but of relatively low sanitary risk score (13, 14, 15, 16) are anomalous results requiring verification by the district supervisor. They may represent remote contamination of the aquifer rather than defects at the tubewell; hence the need for the *urgent resample and reinspection* for confirmation.

Rainwater catchment tanks, like the open dug wells and the converted handpumped tubewells, fall mainly into the intermediate and high risk categories on combined risk assessment (Table 7.6 and Fig. 7.7). There are two principal points of contamination which account for the high proportion of samples which have faecal coliforms: one is the roof catchment area including guttering, the other is the collection point. The latter is typically located at a low point and is poorly protected and drained. For rainwater tanks an equally important consideration is seasonality. The dry season in Java is such that the tanks have been empty for at least 3 months in each of the last 4 years. This underlines their inadequacy to meet the essential service indicator requirements of quantity and continuity of supply.

The most common points of risk of contamination, for all five types of point source systems, are summarized in Chapter 8 on system rehabilitation.

7.3 Peruvian results

7.3.1 Inventory of water supplies

Until December 1985 health region XIII was administered from the regional capital Huancayo and was responsible for six hospital areas, with populations as listed in Table 7.8. The arbitrary classification of 'rural' settlement for communities with less than 2000 inhabitants does not take into account its social and economic structure and production methods, hence the very low proportion of 'rural' population in the area (20 per cent). It is recognized that many other settlements with more than 2000 inhabitants should be considered in this category.

Table 7.8 Populations served by hospital areas in Health Region XIII, Peru (1985)

No.	Name	Total population	Rural population
47	Huancavelica	181,600	22,907
48	Huancayo	549,962	127,523
49	Jauja	189,805	61,950
50	Tarma	119,563	36,593
51	La Merced	280,688	17,441
52	Junin	31,677	5,841
Totals		1,352,395	272,255

Note: Rural population <2000 people.

In 1985 only 231 rural water supplies were registered in the health region by the PNAPR. By August 1987, 404 rural water supply systems had been inventoried and inspected (Table 7.9). The great majority (89 per cent) of these systems were of the simple protected spring sources, although some simple gravity systems took water from unprotected springs or surface sources. Of the remaining systems, 3 per cent were pumped systems and 8 per cent were gravity systems with treatment. Treatment almost invariably comprised sedimentation and slow sand filtration.

These 404 supplies were constructed by various institutions over the preceding 25 years, and although the major input must be attributed to the PNAPR other important construction agencies include councils, regional development corporations and the 'Popular Cooperation' of the Housing Ministry, as well as the communities themselves. The overall rate of construction in the period 1960–88 varied between 4.2 and 13 new systems per year; 16 new constructions were completed during 1986 with the help of investments supported by the US AID (American International Development Agency).

7.3.2 Coverage

Official coverage statistics envisaged that by 1987, 29 per cent of the rural population of Peru would be served by a drinking water supply. Within the country, Junin may be considered as an average department being twelfth (of 29) in rank order of water supply coverage (30.6 per cent) and in the intermediate group (third of 6) for the proportion of the dwellings without piped water supply (with an average of 73 per cent). It is also representative for other indicators such as infant mortality and other socio-economic and health indices.

The Department of Cerro de Pasco, which was formally part of the same health region as Junin, had 27 per cent water supply coverage and was comparable to Junin for most indicators.

In contrast, the Department of Huancavelica was the single worst department for water supply coverage (10.3 per cent) and among the worst of the 'Southern Andes' group for several other parameters.

Rural coverage was estimated by DISABAR to be 54.4 per cent in the pilot region in 1984, but this figure refers to the proportion of rural communities with water supplies rather than the proportion of the total population which is served, and consequently coverage is significantly lower. This discrepancy is illustrated in Table 7.10 for one province in the pilot project region, Tarma.

Cost-efficiency dictates that the population coverage be maximized and therefore that water supplies to larger population centres should be considered first. However, the majority of the rural population is dispersed in many small communities. Thus, the construction programmes have recently placed emphasis on simple gravity systems in small communities.

Coverage within those rural communities with water supply systems was far from complete in 1987. An analysis of the coverage data from the diagnostic study showed that the average coverage in rural communities with water supplies was only 56 per cent. It should be noted that even this figure represented only the proportion of households with a domestic connection or access to a public standpipe. It took no account of water shortages or discontinuity which would be expected to reduce the effective coverage still further, nor of the acceptability or adequacy of the water for human consumption.

Table 7.9 Basic inventory of rural water supplies in the XIIIth health region (1988)

Health No.	Area	Type of system* GST	GCT	BST	Total/area
47	Huancavelica	32	4	0	30
48	Huancayo	132	6	4	142
49	Jauja	81	6	1	88
50	Tarma	71	6	0	77
51	La Merced	28	10	1	39
52	Junin	22	10	0	22
Totals		366	32	6	404

* GST = Gravity without treatment.
GCT = Gravity with treatment.
BST = Pumped without treatment.

Table 7.10 Population size and water supply provision in the province of Tarma: Health Area No. 50 (1987)

	Community			Population		
Size	Total No.	No. Served	Coverage (%)	Total No.	No. Served	Coverage (%)
Rural						
100–499	79	43	54	18,215	7,468	41
500–999	20	19	95	11,303	7,506	67
1,000–2,499	10	9	90	13,130	8,679	66
Urban						
2,500–4,999	4	4	100	16,946	8,134	48
5,000–10,000	2	2	100	16,792	6,213	37
>10,000	1	1	100	46,870	17,342	37
Totals	116	78	67	123,256	55,342	55

Increased coverage is clearly a very high priority. An analysis of water availability showed that water quantity was not a common limiting factor.

7.3.3 Continuity

Analysis of continuity data requires the consideration of two components: daily and seasonal continuity. The surveillance data revealed a significant micro-regional variation in continuity of service. The classes of continuity were as follows:

1. Year-round services from a reliable source with no interruption of flow at the tap. Fortunately the great majority of supplies are of this type.
2. Year-round service with daily variation. The two most common causes of this were: (a) restricted pumping regimes in pumped systems and (b) peak demand exceeding flow capacity of the conduction line or capacity of reservoir.
3. Seasonal service variation due to source fluctuation. This may also have two basic causes: natural reduction of source volume (more commonly encountered in spring sources) and surface source volume limitation due to competition for irrigation use.
4. Compounded daily and annual discontinuity.

The classification reflects broad categories of continuity which are likely to affect hygiene as follows:

1. Daily discontinuity results in low supply pressure and a consequent risk of in-pipe contamination. This also encourages household storage resulting in increased risk of contamination from the generally unchlorinated domestic water. Other associated consequences are lower availability and lower volume use resulting in limitations on washing habits.
2. Seasonal discontinuity forces users to adopt water collection from (usually) inferior and distant sources. A consequence, in addition to the obvious reduction in quality and quantity, is that time is lost in making regular collections.

During the first stages of surveillance (diagnostic phase), 16 per cent of communities had problems of continuity at the time of inspection. Data concerning continuity of service required conscientious interviews and questioning to ensure reliability. Such data gathering was planned for the routine surveillance phase, and a sample is reported in Table 7.11 from a micro-region with copious spring supplies all year round.

7.3.4 Quality

The faecal coliform classification scheme shown in Table 7.2 was applied to grade piped Peruvian supplies in the following way:

Grade A 0/100 ml in all samples in one distribution system on one sampling visit: i.e. conforms to WHO bacteriological guidelines

Grade B 1–10/100 ml in all or any samples on one sampling visit: i.e. low-level, often sporadic, contamination

Grade C 11–50/100 ml in all or any samples on one sampling visit: i.e. medium contamination with significant water-borne disease risk

Grade D >50/100 ml in all or any samples on one sampling visit: i.e. grossly contaminated with high water-borne disease risk

This scheme was used to analyse the water quality data gathered and the results are summarized in Table 7.12. More detailed data are listed in a sample summary which demonstrates the range of values of the principal indicators of level of water supply services (Table 7.11) in a selection of communities.

The full results of the pilot region diagnostic phase confirmed the general findings of the preplan concerning treatment plants. Of the treatment systems, 76 per cent were supplying grossly contaminated water (category D).

Pumped systems also showed a high frequency of contamination with 45 per cent in this category, despite generally superior source water. The rapid deterioration in quality often occurred because of zero internal pressure in the supply lines and distribution network for prolonged periods caused by restricted pumping regimes. This allows the possibility of ingress of contaminants.

In contrast, of the 273 simple gravity systems, which constitute the great majority of the supply capacity, only 17 per cent were grossly contaminated.

7.3.5 Quantity

Analysis of the quantity of water used for domestic purposes proved difficult in the pilot region. The demand for irrigation water in the dry season creates problems of competition with source waters. Also during the 8 dry months, household drinking water connections may be used for irrigation purposes and apparent domestic usage values are often extremely high and almost invariably exceed the 50 litres per person per day (lppd) value for minimum domestic requirements.

This estimate, however, is based on a ratio between measured intake flow at the abstraction point and served population. It does not account for losses through leaks or misuse (like for example household irrigation practices). Under these circumstances, actual water use for hygiene and direct consumption could fall below 50 lppd. In the sample of results presented in Table 7.11 only 3 out of 28 communities fall below the 50 lppd minimum; however, the average daily consumption is >100 lppd.

Despite these high apparent values, continuity and effective coverage may suffer when water supply favours a few houses connected to the low parts of the distribution system. This requires training of the administrative committees and public awareness, and limits the usefulness of the data as indicators of domestic use habits or acceptability of services.

7.3.6 Cost

Despite the obvious need to raise money for operation and maintenance and the publication of recommended tariff levels by the Ministry of Health, tariff collection barely operated until 1985. The actual tariffs collected by the

Table 7.11 Sample summary of surveillance data from the Peruvian Pilot Region (1988). Department of Junin, Province of Huancayo. Surveillance diagnostic of water service indicators

Community	Type of system*	Source	Total population	Coverage (%)	Continuity (% time)	Quality (faecal grade)	Quantity (lppd)†	Tariff cost (I/m)‡
9 de julio	GST	Spring	1590	54	100	B	245	30.2
Achin	GST	Spring	480	75	100	B	54	5.0
Alayo	GST	Spring	925	51	100	B	420	30.2
Andamarca	GST	Spring	690	61	100	B	75	20.0
Andamayo Anexo	GST	Spring	402	21	100	A	580	40.0
Huandar	GST	Spring	705	83	100	D	184	5.0
Chala Nueva	BST	Well	350	43	100	B	123	1.0
Chaquicocha	GCT	Canal	625	77	100	C	124	1.0
Cochapalca	GST	Spring	2500	50	100	D	69	1.0
Cochas	GST	Spring	900	43	100	D	192	4.0
Comas	GST	Spring	1488	53	85	C	1577	0.0
Heroinas Toledo	GST	Spring	695	8	100	B	149	5.0
Hunachar	GST	Spring	705	100	100	D	245	5.0
Huanuco	GST	Spring	210	14	100	A	123	10.0
La Libertad	GST	Spring	380	75	100	D	136	0.0
Matapata	GST	Spring	720	21	100	B	24	20.0

Community	Type of system*	Source	Total population	Coverage (%)	Continuity (% time)	Quality (faecal grade)	Quantity (lppd)†	Tariff cost (I/m)‡
Mrcal. Castilla	GST	Spring	540	51	100	A	160	0.0
Parco	GST	Spring	770	30	100	D	67	5.0
Pomamanta	GST	Spring	918	53	85	B	38	2.0
Pucacocha	GST	Spring	126	60	100	D	69	5.0
Punco	GST	Spring	240	85	85	D	36	0.0
Puquian	GST	Spring	300	58	100	B	202	2.0
Racracalla	GST	Spring	720	53	85	D	60	5.0
S. Fco. Macon	GST	Spring	480	56	100	B	108	2.0
S. José de Quero	GCT	River	720	60	100	D	1200	3.0
S.R. de Huarmita	GST	Spring	750	77	100	B	576	5.0
S. Rosa de Ocopa	GST	Spring	1810	59	100	C	63	5.0
Usibamba	GST	Spring	2000	64	100	A	86	1.0

* GST = Gravity without treatment.
GCT = Gravity with treatment.
BST = Pumped without treatment.

† lppd = Litres per person per day.

‡ I/m = Inti per month for domestic connection.

Table 7.12 Water quality by supply type in the Department of Junin (1987)

System type	Number of supplies	Percentage of systems supplying water of quality:*			
		A	B	C	D
Simple gravity (GST)	273	23	43	17	17
Gravity with treatment (GCT)	25	4	8	12	76
Pumped (BST)	9	11	22	22	45
Total	307	21	39	17	23

* Faecal coliform grades.

administrative committees and the current Ministry of Health recommendations are presented in Table 7.13.

There have been several major agencies concerned with rural water supply construction in the pilot region (Table 7.14). The greatest activity is found within DISABAR, SENAPA, COOPOP (Popular Cooperation), CORDE (the Regional Development Corporation) and the municipalities. In addition, some of the systems were financed and constructed by the communities themselves. It is possible that other agencies may also have been involved in construction. However, those listed are the ones which were defined by members of water surveillance teams in the pilot region when constructing their inventories. It may be noted that as the official lead agency in rural water supply DISABAR is the only one with a clearly defined tariff policy which is beginning to work financially and administratively.

7.3.7 Administration

It is mandatory for interventions by DISABAR that any community receiving a water supply system should form an administrative committee. This committee receives statute book and must organize tariff collection, and operate and maintain the system. The supervision of these activities is the responsibility of the DISABAR Programme. In contrast, systems constructed by CORDE and COOPOP are handed over absolutely to the communities and those agencies had no involvement whatsoever in their subsequent administration, operation and maintenance.

The indicators used to assess the success of administration arrangements were as follows:

(a) the proportion of villages with an administrative committee;
(b) the proportion of systems with operators;
(c) the proportion of these operators who had received training;
(d) the operation of system with an existing tariff.

Table 7.13 Tariff structure by type of water supply system in the Department of Junin

System type	MOH recommended minimum tariff (Intis)	Percentage of communities paying tariffs of:					
		0–4.9	5–9.9	10–14.9	15–19.9	20–29.9	>30
		(Intis/family/month)					
Simple gravity	15	86	7	4	<1	2	—
Gravity with treatment	25	74	26	<1	—	<1	—
Pumped	30	55	33	—	11	—	—

Notes: Exchange rate 50 Intis = $US1 (1987).
Per capita income unqualified labourer = 50 Intis/day.

Table 7.14 Contributions of various construction and administrative agencies to water supply systems of the pilot region

Construction agency	Number of systems	Percentage of all constructions
DISABAR	215	70
SENAPA	4	1
Municipality	3	1
Self-funded	23	7
COOPOP	10	3
Others/unknown	52	17

These indicators are presented in Table 7.15.

Failures in administration could usually be attributed to a lack of training of committees. The surveillance teams were asked to evaluate the interest of each village in collaborating in the improvement of their water supply system during the diagnostic phase. Overall, 97 per cent of communities were interested in collaboration by providing manpower, and 39 per cent were also able to offer financial support.

These data imply that with well-orientated technical support self-help programmes of rehabilitation, repair and operation by surveillance of water quality and water services then becomes a very important dimension in the efficient direction and application of resources.

7.3.8 Sanitary risk assessment

It is obvious that piped supplies are structurally far more complex than point sources of drinking water. Consequently, the sanitary inspection report forms

Table 7.15 Basic indicators of water supply administration for rural communities

Indicator	Percentage of communities
Administrative committee	93
Operator	67
With tools	43
Trained	43
Tariff collection	71

are much more detailed with many more potential points of contamination to examine for and to sample at. The Peruvian surveillance report forms ran to 10 or 12 pages depending on whether the system had a treatment plant or not. As demonstrated in Chapter 5, it is valuable to separate out basic data from source risks and subsequent risks in treatment, reservoirs and distribution by separate sections in the report form. These risks have all been summarized for a complete community system with treatment in Table 7.16 to provide a sanitary inspection risk index.

Questions were designed so that all 'yes' answers indicated that the system was vulnerable to contamination and thus a risk to public health. The risk score is therefore obtained from the sum of the number of 'yes' answers. The total possible 'yes' answers are '20' which would represent highest risk.

When the risk index is compared with the level of faecal pollution we should not necessarily expect to find a correlation. What we should aim for are sufficiently thorough sanitary inspections to reveal the key points of risk and thus provide a general overview of the operational status of the system. Even though the system is at risk it does not follow that it will be contaminated at each point of risk whenever we analyse a sample from that point. Analysis alone will therefore often underestimate the risks. On the other hand there may be occasions when the point of contamination is hidden from inspection, e.g. in the distribution system underground, in which case we would hope to detect such contamination and identify the source of it by a systematic sampling of the main branches of the distribution mains. The sanitary inspection risk index has been plotted against the A, B, C, D grade of faecal coliform contamination for samples on one sampling day for over 100 systems in Fig. 7.8. This demonstrated the tendency for faecal bacteriological grade to be broadly correlated with sanitary inspection risk index in spite of the fact that the risk index has not been weighted in favour of points of potential high level contamination. Five systems with treatment in Fig. 7.8 are denoted by inverted commas, '60', and they all have particularly high level faecal contamination 'D' in the presence of high level sanitary inspection risk. All of these five were surface-water sources with non-functional treatment plants incorporating sedimentation and sand filtration. The diagnostic survey subsequently demonstrated that the rural treatment plants have uniformly failed to reduce the contamination of surface-

Table 7.16 Sanitary survey form for the assessment of risks of contamination of drinking water supplies. [See also Figure 8.2].

I Type of facility **GRAVITY FEED PIPED SUPPLIES WITH TREATMENT**

General information :

1. Location : Health centre ....
   : Village ....
2. Code No. ....
3. Water Authority/Community Representative signature ....
4. Date of visit .... Signature of sanitarian ....
5. Is water sample taken? ....Sample No. ....Faecal coliform grade ....

II Specific diagnostic information for assessment

		Risk	
		Yes	No
Abstraction			
1.	Is the surface water inlet screen absent, damaged or blocked?	☐	☐
2.	Is the minimum head device missing or not working?	☐	☐
3.	Is the flow control (V-notch) non-functional or missing?	☐	☐
Treatment			
4.	Is the pretreatment tank (sedimenter) dirty, or are the wash valves not operating?	☐	☐
5.	Does the sedimenter lack adequate baffles, or is retention time less than 2 hours?	☐	☐
6.	Are the slow sand filters blocked or bypassed?	☐	☐
7.	Is the depth of sand in the filters less than 60 cm?	☐	☐
8.	Is either the flow control at the inlet or outlet of the filters not working?	☐	☐
9.	Does the plant lack facilities for washing filter sand?	☐	☐
10.	Is the water leaving the plant greater than 5 TU?	☐	☐
Conduction pipe to reservoir			
1.	Is there any point of pipe leakage between the source and the reservoir?	☐	☐
2.	If there are any pressure break boxes, are their covers insanitary?	☐	☐
3.	Is the inspection cover on the reservoir insanitary?	☐	☐
4.	Are any air vents insanitary?	☐	☐
5.	Do the roof and walls of the reservoir allow any water to enter?	☐	☐
6.	Is the reservoir water unchlorinated?	☐	☐
Distribution pipes			
7.	Does the water entering the distribution pipes have less than 0.4 ppm free residual chlorine (<0.4 mg/l)?	☐	☐
8.	Are there any leaks in any part of the distribution system?	☐	☐
9.	Is pressure low in any part of the distribution system?	☐	☐
10.	Does any sample of water in the principal distribution pipes have less than 0.2 ppm free residual chlorine?	☐	☐

Sanitary inspection risk index = total score of risks ..../20

Signature of sanitarian ....

water source and present the highest public health risk. The reasons for these failures were identified and strategies for improvement have been developed as summarized in Chapter 8 on rehabilitation.

7.4 Zambian results

7.4.1 Inventories and coverage

By the end of the pilot project there was no full inventory or classification of water resources in Mongu district held either at provincial or district level in the Ministry of Health. Furthermore, there were no formal inventories of local water sources held at health centre level. This is partly because many traditional sources are seasonal and change frequently. In contrast, all permanent protected sources were listed in the DWA inventories. Traditional and non-permanent sources were not listed, but most of those used for potable purposes were known to health assistants. By examining data and information from a variety of sources, estimates of water supply coverage in the district could be made.

In 1986/87 the population of Mongu district was approximately 120,000 of which around 80,000 were rural. Reports presented by the health assistants showed that the project covered more than 50 per cent of the population of the district, although reports from Iloke and Sitoya were missing (Table 7.17). Due to long distances and bad roads as well as lack of transport, only those villages within 5–10 km of the rural health centres were visited.

Written reports provided by the health assistants showed that the main water sources comprised traditional unprotected wells, while protected water sources represented only a minor proportion of the total number of supplies (Table 7.17). The townships were supplied with piped groundwater, while three small communities were supplied with piped water from canals or streams. It should be noted that in terms of populations, protected sources were thought to be serving an equal population to the traditional sources. This was based on health assistants' lists which covered approximately 50 per cent of protected sources which were estimated to serve about 45 per cent of the total population.

Traditional water-holes are unprotected sources from which water is scooped out by a bucket. The water-level is typically only 0.5 m below the soil surface, i.e. within arm's length. Most of these traditional sources serve significantly less than 100 people but no systematic survey of numbers using a source was carried out.

In the case of new installations only rough estimates were available as to the population served. These estimates were derived from the fact that to qualify for an improved source one of the principal selection criteria for construction of boreholes and dug wells was that there should be more than 30 households or 180 people.

No firm conclusions can therefore be drawn about the level of coverage by different types of facility throughout the district since data on the population served by each type were not collected systematically.

Sanitary inspection risk score

Faecal coliform grading	0	1	2	3	4	5	6	7	8	9	10	11	12	13	14	15	16	17-20
D			50.	13,	3,			49, 28,	'60' 51, '52' 33, 81,82	'29' 18, 75, 54, 70,	'86' 62, '71'	56, 79, 80,	69,	32, 42,	2,	102,		
C				26, 59,		68, 105,	11, 107, 163,	24, 111,	4, 8, 40, 46,	65, 94, 110,	38, 58, 98,		19, 103, 71,	44, 91, 47, 64,				
B							30, 72, 77, 99,	16, 20, 21, 23, 37, 55,	6, 10, 17, 84,	93, 1 104, 113, 31, 41, 66,	88, 95, 106, 139,	53, 67, 83, 100,	34, 7,	25, 57, 90,				
A		119, 142, 147,		122,		134,	5, 14, 45, 112,	136, 22, 63,	9, 12, 43, 73,	74, 85, 97, 101,			35,	39,				

No risk No action	Low risk	Intermediate to high risk	Very high risk: Urgent action

FIG. 7.8 Combined risk analysis of sanitary inspection and faecal coliform contamination of gravity fed piped water supplies. Survey period 1986/87, study area Junin, Peru. Code: Number = community; '60' = 'treated' supply, sedimentation and filters. All others are untreated.

Table 7.17 Inventory of population and numbers of water sources covered by the project in Mongu district, Zambia

Health centre	Population	Handpumps	Protected wells	Traditional with windlasses	Canal/ water-holes	Protected stream	Springs water-hole	Piped supply
1. Ikwichi	4,068	12	10	149	ND	ND	ND	0
2. Kalundwana	4,200	6	9	>50	ND	ND	ND	0
3. Limulunga	10,000	ND	10	36	ND	ND	ND	1
4. Luandui	3,641	7	5	38	ND	ND	ND	0
5. Lukweta	4,443	2	ND	200	ND	10	ND	0
6. Mongu (urban)	>29,000	0	0	0	0	0	0	1
7. Mongu (rural)	>13,000	ND	ND	ND	ND	ND	ND	1
8. Nalikwanda	8,000	ND	17	122	ND	ND	ND	0
9. Nangula	4,500	1	5	197	1	ND	ND	0
10. Ndanda	1,668	ND	3	>1	>1	ND	ND	0
11. Safula	5,000	12	6	191	ND	ND	ND	1
12. Ushaa	3,865	1	15	90	ND	ND	ND	0
13. Iloke	—	—	—	—	—	—	—	—
14. Sitoya	—	—	—	—	—	—	—	—

ND = Not surveyed.
— = Records missing.

7.4.2 *Quantity and continuity*

The majority of the urban population of Mongu district was served by piped water. In Mongu township the continuity of this service was poor, resulting in water shortages during the day, especially in the dry season. This discontinuity did not apparently result in any deterioration in water quality. The shortage of water was mainly due to leakage through pipes, valves, taps, etc. estimated to result in a loss of 65 per cent of treated water. It is not known if there were water shortages in other towns.

The rural water supplies were wells, streams and traditional wells. The boreholes and shallow wells provided reliable supplies and were installed after 1980. In some sources the water-level became low towards the end of the dry season and water shortages occurred. There were no readily available estimates of the amount of water drawn from these sources, but water carried back to houses averages around 10–12 lppd, according to DWA surveys. Although this is well below the 50 lppd target in Peru, these amounts suffice for drinking and cooking purposes, and it was suggested that most people wash at sources which are distinct from those used for drinking purposes.

7.4.3 *Administration and tariffs*

No numerical data are available on local, community-level administration of water supply although there was a strong commitment to education and community participation by DWA/WASHE. This involved cooperation with whichever extension workers were available in the locality of the training in the water supply improvement (e.g. health assistants and agricultural training in community participation and good water use), and extended to the provision of spare parts and bleaching fluid for purchase by committees. Education of the community, the organization of tariffs and the training of individuals in maintenance for new supplies was largely CEP/WASHE activity, but health assistants were planned to be the main agents in following up after construction. WASHE health education included training in good practice in water use, domestic hygiene and oral rehydration therapy.

It may be concluded that with a preponderance of point source systems this type of water supply strategy and dispersed population does not lend itself to the development of a tariff structure and is dependent on charity and self-help.

7.4.4 *Quality*

About 600 water samples were collected and analysed for total and faecal coliforms. Of these, 226 samples were collected by the health assistants for the project, who concentrated on spot samples of piped water and traditional sources. The remaining samples were collected by the DWA, who concentrated on more regular monitoring of handpumps and shallow wells.

A summary of results obtained for faecal coliform analysis is shown in Table 7.18. This table gives the percentage of samples from different sources graded A, B, C and greater.

From the basic figures in Table 7.18, it is apparent that water from boreholes, handpumps and standpipes give good bacteriological water quality. Shallow wells provide lower quality, but are seldom highly polluted. Traditional sources are significantly more polluted and one in three samples showed relatively high levels of pollution. However, it should be noted that in Mongu district the level of contamination for both shallow wells and traditional unprotected sources is remarkably low compared with that found in the other pilot projects and by comparison with neighbouring countries in Africa. It has been suggested by Sutton (personal communication) that this low level of faecal contamination may be attributed to:

(a) extremely sparse human and animal population density;
(b) the distance between human habitation and the water source which is generally more than 100 m away;
(c) the Kalahari sand in Mongu which, in contrast to the fissured limestone of Java, provides effective bacteriological filtration of groundwater in shallow unprotected water-holes.

Against these points it should be recorded that there were major sampling problems and no quality assurance carried out independently on the analysis.

The water samples were transported from the source to the laboratory in a kit provided by WHO. This kit consisted of a wooden box containing a copper container for cooling water and four compartments for 250 ml sterile water-sample bottles. The kit was heavy and inconvenient to carry, and did not keep the sample bottles cold under the conditions in Mongu.

Many samples collected did not arrive at Mongu laboratory because the health assistants had to pay for transport to get to the laboratory. Many of those which did arrive had been in transit for 48 hours and therefore would have at least a one log reduction of faecal counts due to exposure to high ambient temperatures. All the analyses were done in the Mongu laboratory and there was limited cross-checking by an independent consultant but none by a central reference laboratory. Under these circumstances it would have been more effective to have used the field portable test kits used in the other two projects.

7.4.5 Sanitary inspection and risk assessment

No systematic sanitary inspection procedure was applied consistently by the health assistants throughout the project and therefore no quantitative risk assessment data are available. However, the DWA carried out annual inspections of all protected sources, to assess their water condition, reliability and the level of activity of the village water committees. These inspections included

Table 7.18 Percentage of water samples from different sources with faecal coliform (FC) graded according to level of contamination; results from several countries

Source	No. of sites	No. of samples	Faecal coliform grade* A	B	>C
Zambia, Mongu (Utkilen and Sutton 1989)					
1. Handpumped tubewell (>10 m)	32	40	100	0	0
2. Standpipe from borehole	34	35	100	0	0
3. Handpumped dug well (concrete rings)	7	14	93	0	7
4. Dug well (windlass)	98	266	78	14	8
5. Traditional source (unprotected)	108	148	42	20	38
6. Spring	4	9	89	11	0
7. Stream	6	14	28	29	43
8. Unprotected shallow well	14	17	53	41	6
Malawi (Lewis and Chilton 1984, 1989)					
1. Handpumped tubewell and boreholes		300	60	34	6
3. Handpumped dug wells			81	14	5
4. Unprotected dug wells		60	8	2	90
Java, Gunung Kidul					
1. Handpumped tubewell					
(a) Deep (>10 m)		44	59	25	16
(b) Shallow (<10 m)		100	72	14	14
3. Handpumped 'protected' dug well		97	20	32	48
4. Open dug well with parapet		55	5	54	34

* Faecal coliform grading: A = 0/100 ml; B = 1–10/100 ml; C = 11+/100 ml.

comments on state of all parts of the well from lifting device to slabs and drainage and were done by people independently from the project sampling.

In October 1986, before the water quality control project started, an inspection was carried out of 61 shallow dug wells which were subsequently sampled by the project. Of these 35 were later found to have no faecal coliforms and 10 had less than 10 faecal coliforms/100 ml. Of the 26 which had some faecal coliforms: 3 had been identified as being at risk as they were new wells for which surrounding slabs had still to be constructed. All had <10 faecal coliforms/100 ml, i.e. grade B, and 22 had been identified as being at risk for one or more of the following reasons:

1. they were drying up, so turnover of water was very low;
2. water was turbid and needed bailing;
3. slabs were badly cracked or broken or there were gaps between rings;
4. consumers had made complaints about quality.

7.5 Comparative evaluation of results

7.5.1 Inventories and coverage

In Indonesia and Zambia after the initial selection of the target area and villages there was no continuous process for updating inventories of traditional sources. In the case of Indonesia this was partly because they were so numerous and in the case of Zambia because many traditional sources are no more than holes in the ground which dry up in the dry season. In Peru, however, the development of inventories was seen as an on-going activity and from the beginning of the programme data bases for basic inventories were established for all communities within the project region. These inventories were held by the regional surveillance office (Huancayo) and reported at least annually to the national surveillance team along with copies of all surveillance reports which include both inspections and analysis. In Indonesia the development of water source inventories has now been adopted in Sumatra as a result of a WHO/UNDP programme initiative (project INO CWS 007) 'The Institutional Development of Water Supply and Sanitation in Bengkulu and Lampung Provinces'. The magnitude of the problem is indicated by the estimate of Lloyd (1988a) that in Lampung alone there are between half and three-quarters of a million dug wells. He also suggested that an additional indicator of coverage for point source systems is required which is distinct from that for piped supplies. At village level coverage may be defined as the percentage with access to an improved source. Thus the value is derived from the average number of people using each type of facility. In many villages in Indonesia this value is between 4.5 and 7 per dug well, implying that almost every family has a private traditional dug well. This may therefore be considered as the most basic level of water supply service. As we have already seen in section 7.2, the bacteriological quality of this water is generally very poor. It should also be noted, however, that community dug wells with handpumps serving upwards of 30 people are not necessarily an improvement if this implies a greater distance for people to walk to obtain water of a similarly poor quality. However, in the case of Zambia (section 7.4) and Malawi it was demonstrated that dug wells could provide safer water if the dug well was properly completed.

It is clear that major improvements in water supply service levels can only be achieved for nuclear and ribbon distributed population centres because they are susceptible to piped supply development. In Indonesia piped supplies are generally limited to such sources when the population exceeds several thousand. The development of piped supplies is then the responsibility of a transitional water authority and not the Ministry of Health environmental health division. This is therefore a major step towards professional management and away from community involvement. Even within these nuclear and ribbon community types the coverage is at present typically less than 50 per cent and in many provincial capitals it is still below 30 per cent with tap connections. Probably about 80 per cent of all Indonesians depend on traditional dug wells as their principal source of domestic water.

It was thus far from clear from Gunung Kidul data how to express this most essential service indicator, coverage, which is vital for government to be able to plan their water supply improvement strategies. The Zambian project is equally unhelpful in this respect, and it is the WHO/UNDP Sumatra project which is leading the way in providing detailed maps and listings of all point source supplies in advance of borehole exploitation (Lloyd 1988).

By contrast with Indonesia and Zambia, the Peruvian rural water supply strategy has focused on numerically, geographically, socially and administratively well-defined target communities which lend themselves to the installation of a single piped water supply system. This is a most obvious and major advance over multiple unpiped point sources, and from the surveillance viewpoint has major advantages in being susceptible to classifications using the critical indicators of water supply service.

From the start of the Peruvian pilot project a principal objective was the systematic collection of basic data by the surveillance teams. Using report forms revised after the preplan pilot project the data were at first held in traditional filing cabinet archives. By the end of 1985 a data base was developed using OMNIS 3 for Apple McIntosh computers and national level staff trained to use this at DITESA in Lima. This was subsequently translated to DBase for IBM compatible machines.

It was demonstrated in section 7.3 that useful data, including coverage, were obtained for all communities surveyed and a sample was displayed in Tables 7.10 and 7.11.

7.5.2 Continuity and quantity

Table 7.11 demonstrated that in Huancayo, Peru, in that particular micro-region continuity was not a problem. Elsewhere in the country, even in the pilot project area, continuity of supplies may be a critical problem arising from climatic changes and geologically unstable conditions. Seismic phenomena sometimes terminate a source for ever and this is also the case in Indonesia. However, the commonest phenomenon is the problem of seasonality of rainfall in all three project areas. It has been pointed out that the dry season results in the disappearance of traditional water-holes in Zambia, the abandonment of rainwater tanks in Java and the reduction of spring and stream flow in Peru. On the other hand, water quality tends to improve in the dry season. This fact may, however, be insignificant if the overwhelming problem is lack of continuity due to shortages and breakdown in simple technology. It has been reported from many countries (Gibbs 1983; Tschannerl and Bryan 1985) that the failure rate of handpumps for example is greater than 30 per cent in the first year. Inevitably people then return to poorer traditional sources.

With respect to both continuity and quantity the Peruvian pilot region has much more favourable conditions than either of the other two project areas: its continuity of supplies is generally very high and the amount of water available at the tap is greater than 100 lppd. By contrast, the Indonesian study area suffered from drought and under emergency conditions people had either to walk several kilometres to underground stream outlets in some areas, or depended on tankered water when dug wells also dried out. Because of the water carriage problem from point source supplies it is, however, very likely that both in Java and in Zambia the quantity of water used in the home is less than a quarter of that available in the Peruvian study area, even under normal weather conditions.

7.5.3 Quality

The use of the faecal coliforms grading scheme (A, B, C, D, E) makes the comparison of data from the different projects relatively straightforward. But before making such comparisons it should be noted that there were significant differences in the way that coliform bacteriology was carried out.

It has already been pointed out that sampling and sample transport procedures in Zambia were deficient and that this could seriously affect the results and the grading. In Indonesia sample transit time was 0–3 hours. In Peru it was 0 hours, because processing and incubation of the sample were carried out on site, whereas in Zambia samples were often in transit, unrefrigerated, for up to 2 days.

On arrival in the Mongu laboratory the water samples were analysed by the laboratory technician who was trained to do bacteriological water analyses by both the multiple tube (MPN) method and the membrane filtration (MF) method according to the WHO Guidelines volume 3. Early work relied on the MPN method but later the MF method was adopted on the advice of WHO consultants.

The Indonesian pilot project also used both standard methods whereas the Peruvian project relied exclusively on the MF technique.

The case for using the faecal coliform test in tropical countries was presented in section 6.1 but it is also worth comparing the two standard methods to assess their merits and appropriateness. From Table 7.19 it is very clear that the MF method has major advantages over the traditional MPN method. It should also be noted that in many countries the MPN method is carried out incorrectly partly because there are more stages to do before a confirmed result can be correctly reported but also because some countries are adopting procedures, such as the H_2S test, which have not been properly evaluated. This has led to a number of countries in the South-East Asia and Pacific area reporting high levels of false positive coliform results. Hewison *et al.* (1989) reporting specifically on the evaluation of the H_2S test in north Thailand concluded that 'the test cannot be used with confidence to test the bacteriological quality of water'. There is thus a major risk that many rural supplies are being condemned on specious but invalid evidence. We therefore recommend that if the MPN method has to be used in rural water supply investigations it should:

(a) be used with traditional standard coliform selective media and standard procedures;

Table 7.19 Comparison of methods for coliform and faecal coliform analysis

Factor	Membrane filtration	Most probable number
1. Precision and reliability	Precise colony count. 100 ml tested	Large error, +300% to −30%. Some labs use only 50 ml
2. Ease of execution	Simple. Robust media. Selective lauryl sulphate broth recommended for long shelf life	Complicated. Errors common. Many labs use non-selective lactose broth and do not confirm the presumptive count
3. Speed	*E. coli* + separate coliform result in 14 hours if required	Coliform result in 2 days. Confirmed result in 3 days
4. Principal equipment	Small amounts of Petri dishes and media (3 ml/test)	Large amounts of media and glassware required.
5. Costs	Pressure cooker ($US 50–100). Membrane, pad + media 10 US cents*	Autoclave ($US 1000–2000). Media 10 US cents/test

* UK prices 1989.

(b) with properly monitored incubation temperature (37 °C); and
(c) that the test is properly completed to confirm the presence and level of faecal coliform contamination.

If this advice is not followed and no quality assurance is ever carried out by a reference laboratory, then the data should be disregarded and sanitary survey results used instead.

In the case of our own Zambian data we were satisfied that the laboratory procedures were adequate from spot checks by WHO consultants, but most concerned about sampling procedures. Apart from the protracted sample transport time, the practice of using cotton wool plugs (also in Indonesia) is not recommended. Consultants observed poor hygiene in handling of plugs which were not infrequently dropped before being returned to the sample bottle. The use of a protected screw cap (as illustrated in section 6.3.2) is strongly recommended. Supervision of sampling and inspection was also a weak point in Zambia and Indonesia. In spite of these weaknesses the bacteriological results generally make good sense.

Examination of the A, B, C graded faecal coliform data in Table 4.18 shows that 100 per cent of tubewells and standpipe samples from boreholes comply with WHO guideline value of zero, i.e. grade A. This is in marked contrast to both the Javanese data and the Malawi data of Lewis and Chilton (1989) which shows 59 and 60 per cent grade A respectively for deep tubewells and boreholes. It had been suggested that this major difference may be due in part to different geology, but our suspicions were aroused initially by the discrepancies between the quality of dug well water. Again the Zambian data indicate a remarkably good quality with 93 per cent of handpumped and 78 per cent of windlass systems providing grade A water. By contrast the Javanese handpumped wells produce only 20 per cent grade A, while only 5 per cent of open dug wells are grade A. Likewise only 8 per cent of the Malawi dug wells are grade A. In another study by Lewis and Chilton (1984) backfilling of dug wells to ground level and use of porous concrete rings improved protection so that their handpumped dug wells were 81 per cent grade A in the dry season. These results are much closer to the comparable Zambian data for dug wells which are also protected by concrete rings. They serve to emphasize that dug wells can provide acceptable water quality if sanitary completion is good. In Indonesia sanitary completion is very poor and the great majority are lined with open stonework.

In section 7.2 it was noted that the Indonesian dug wells were the most high risk type of point source supply, both from the point of view of faecal coliform grade (D/E) and as a result of combined risk assessment. Inspection of Table 7.6 revealed that 22 per cent of open dug wells were graded D/E and that 20 per cent of converted handpumped dug wells were still D/E. The conversion had, however, transferred 9 per cent out of the E category but this improvement is marginal compared with Zambian and Malawian systems of the same type (Table 7.18). These basic differences provide a clear pointer to the remedial action and improvement strategy outlined in Chapter 8.

The quality data for the Peruvian piped systems cannot be compared with the other two pilot projects but none the less the faecal coliform grading scheme still provides a similarly useful basis for improvement. The one common feature, however, with each of the projects was the complete failure of their respective institutions to establish routine disinfection control at rural level.

7.6 Control requirements

There is a strong case for introducing disinfection into rural water supply practice at least for piped supplies. Not only does a free residual of chlorine provide protection against the introduction of many pathogens in a distribution system; it may also persist long enough in a bucket of water drawn from a standpipe system to control post-collection contamination. For example, an initial free residual of 0.6 mg/l was reduced by 66 per cent to a still useful 0.2 mg/l after 12 hours at 21 °C. This is particularly important in view of the commonly reported observation that an initial level of 0–10 faecal coliforms per 100 ml standpipe sample is converted to 100–500 faecal coliforms per 100 ml in unchlorinated water, through careless transport, storage and use.

There have been very few controlled experiments on the reduction of the incidence of infant and early childhood diarrhoea resulting from chlorinated versus non-chlorinated water. However, an unpublished report from West Bengal (Institute of Child Health 1982) examined the influence of regular cleaning and refilling of household water pots with chlorinated water. A control group received a placebo of water which was unchlorinated. The reduction of childhood diarrhoea in the group receiving chlorine was 75 per cent greater than in the placebo group. If true, this provides as strong a case for promoting chlorination as for oral rehydration. In the meantime less stringent guidelines for rural water supply quality have been proposed. Volume 1 of the WHO Guidelines (1984/85) state that for unpiped supplies: 'the objective should be to reduce the coliform count to less than 10/100 ml, but more importantly to ensure the absence of faecal coliform organisms . . . Greater use should be made of protected groundwater sources and rainwater catchment, as these are more likely to meet the guidelines for potable water quality.'

Feachem *et al.* (1978) have examined diarrhoeal and typhoid reporting to hospitals in Lesotho in areas where water supply has been improved and others which have not. They concluded that where water quality was the only parameter improved and where there was neither any increase in the volume of water used nor any change in domestic and personal hygiene, then there was no detectable change in the incidence of water-related diseases in study villages. Furthermore, they concluded that the spatial and temporal distribution of typhoid reportings did not support a water-borne transmission hypothesis. They also believed that the wet season peaks of diarrhoeal disease were not associated with poor water quality. They suggested that the survival of pathogens in the moist, warm conditions of the wet season in Lesotho may be significantly better than in the dry, cold conditions of the winter, and that this might be an important factor in food contamination and hence food-borne transmission. However, they also reported that 69 per cent of water sources tested showed bacterial contamination to have overall mean concentrations 5.4 times higher in the wet than in the dry season. Although they rejected a water-borne explanation of both typhoid and diarrhoea they were careful to point out that the data at their disposal were insufficient to make other than informed guesses.

The WHO pilot projects and the Lesotho data demonstrate that in the rural environment simple improvement projects such as the protection of spring water can readily reduce the level of faecal coliform contamination by almost two orders of magnitude, from say 1000 (grade E) down to 20 faecal coliforms/100 ml (grade C). But these values represent average reductions and there is no guarantee that any source will remain with low levels of contamination throughout the year, and since there is no barrier against contamination other than source protection their safety is highly suspect. Feachem *et al.* (1978) argue that the improved supplies are much better than unimproved supplies and that since diarrhoeal disease and typhoid do not appear to be primarily water-borne, the costs and maintenance implication seems to be that the regular occurrence of low levels of faecal coliforms is acceptable. It is certainly inevitable unless there is an additional measure of protection such as routine chlorination to provide a free residual of hypochlorous acid in the distribution system. Feachem *et al.* condone the Lesotho village water supply policy which is not to undertake disinfection, because of the problems of operation and maintenance. This is not an argument with which we concur. While it must be admitted that the training needs for operating and maintaining water supplies and sanitation facilities in the developing countries present the greatest difficulty it is generally agreed that this sector should have a high priority. In the case of the Andean group, DelAgua in Peru and CINARA in Colombia have been actively promoting the development of small-scale treatment including multistage gravel and sand filtration as well as terminal disinfection. These schemes routinely reduce D/E grade water to A/B grade.

It is theoretically possible to relate the intensity of faecal indicator pollution to the probability that water-borne disease will occur. In practical terms, however, the variation in response to ingestion of known pathogens is such that those involved in setting guidelines prefer to set a conservative guideline which is believed to give a wide margin of safety. This is at present not only practically unattainable in most rural supplies but possibly less relevant in sparsely populated areas where coliform contamination, at least, may be dissociated from faecal contamination. It is extremely rare that a known level of faecal pollution is associated with an outbreak of water-borne disease. An exception was Montrose in Scotland where the failure of disinfection on a river-derived supply resulted in about 50 per cent of the town's population suffering a mixture of viral gastroenteritis and *Shigella sonnei* dysentery (Green *et al.* 1968). In this particular case the faecal coliform count in distributed water was between 50 and 100/100 ml, i.e. grade D. This is a reasonable justification for grading such water at very high risk.

7.7 Health implications of results

Very few developing countries have developed a systematic approach to disease data gathering which is consistently applied down to health centre and health post level. In rural areas the recovery of epidemiological data presents particular problems in terms of obtaining reliable data from statistically adequate populations. In terms of the relationship between water-related disease and water supply service indicators the problem is compounded by the fact that even if the health data recovery system is in place the geographical distribution of the human population rarely matches the area served by a water supply. None of the pilot projects was able to provide adequate disease statistics.

In Table 7.20 the key indicators of service level have been summarized to provide a general qualitative assessment of the health implications of the results for each pilot project area. Population density has also been added since it not only influences the level of contamination of the natural environment but also affects the pressure on financial resources. The best we can do with this type of information is a prediction of relative risk. This is qualitative because the data have only been collected quantitatively for Peru.

Although one of the roles of the surveillance agency is to monitor and control water-borne and water-related disease outbreaks, the primary role is more far reaching than fire-fighting the latest outbreak by issuing for example 'boil all drinking water' notices. The surveillance planner and coordinator must look

Table 7.20 General assessment of health risk associated with water supply service indicators in the three pilot project areas

Service indicator	Indonesia (Gunung Kidul)	Zambia (Mongu)	Peru (Huancayo)
1. Piped coverage	<20%	<20%	>50%
2. Continuity	Poor	Poor	>85%
3. Quantity	<30 lppd	<30 lppd	<100 lppd
4. Quality	Bad	Good	Intermediate
5. Sanitary risk	High risk	Intermediate to high risk	Intermediate to high risk
6. Population density	High	Low	Low
Predicted associated health risk	Very high risk	Intermediate to high risk	Intermediate to low risk

beyond the day-to-day problems and begin to develop an infrastructure to tackle the causes of these problems and prevent them. To this end we have used the Peruvian pilot project to provide data bases to identify and quantify the risks to which communities are exposed. In this respect the water supply service indicators are more useful than disease statistics. The service indicators not only indicate health risk exposure but also indicate the remedial action which is required to reduce risk.

CHAPTER 8

Remedial action

8.1 Remedial action strategy

The most important outcome of the risk assessment approach is the estimate of relative level of risk for a collection of sources and systems. This estimates the urgency of need for remedial action.

The procedures described in previous steps were applied in pilot projects for the first time in the period 1985–88. It has been demonstrated that the graphical plot provides the basis of *a strategy for prioritizing remedial action* by classifying each facility into one of four levels of action:

1. very high risk and hence most urgent remedial action;
2. intermediate to high risk requiring action as soon as resources permit;
3. low risk, lower priority for action;
4. no risk, no action.

The objectives of the strategy are readily summarized in Fig. 8.1. This emphasizes, by means of the arrows, that the combined risks should be progressively eliminated by transferring systems from top right to bottom left as a result of improved protection and rehabilitation. Once the surveillance office has established a routine reporting programme, the risk zone scores can be tabulated for each source type as classified in the original sanitary survey. A priority ranking for remedial action can then be established which also takes into account the populations supplied.

8.2 Indonesian pilot project

In Gunung Kidul it has been demonstrated that the combined risk assessment approach can readily identify sources requiring urgent action. The sanitary inspection data which have been used to produce this assessment may also therefore be used to identify the most common points of risk (Lloyd 1988b). These have been listed in rank order in Table 8.1 for each of the main drinking water sources. With more experience and data in the future we will be able to correlate the most important points of risk with bacteriological contamination. At this stage, however, it seems logical to remedy as many of the points of risk as are affordable.

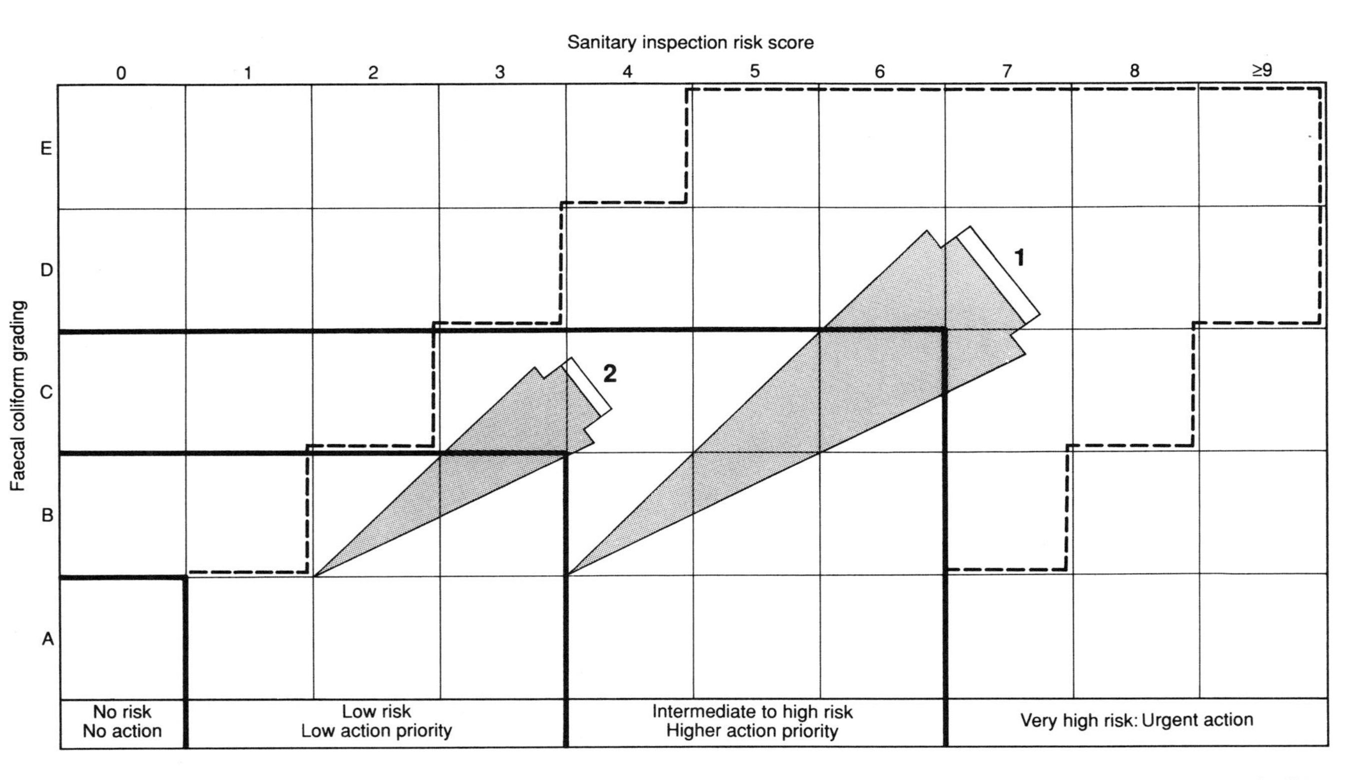

FIG. 8.1 Remedial action strategy using the combined risk analysis of sanitary inspection and faecal coliform grading of drinking water facilities to show the required improvement trends. (1) = first priority improvement from very high risk to low risk; (2) = second priority improvement from intermediate to low risk. The objective is to improve simultaneously bacteriological quality and to reduce sanitary risk score.

Table 8.1 Rank order of commonest points of risk of pollution of point sources detected by sanitary surveys in Java (Lloyd 1988)

Rank order	Open dug well	(%)	Covered dug well + handpump	(%)
1.	Drainage pipe too short (<3 m) or faulty	73	Drainage pipe too short (<3 m) or faulty	71
2.	Ponding on plinth	66	Polluted surface water within 10 m of well	65
3.	Plinth cracked	64	Well lining not rendered to 3 m	62
4.	Well lining not rendered to 3 m	59	Ponding within 2 m of well	60
5.	Polluted surface water within 10 m	55	Ponding on plinth	48

Rank order	Shallow tubewells + handpump	(%)	Rainwater catchment + tank	(%)
1.	Drainage pipe too short or faulty	77	Insanitary collection point	70
2.	Plinth cracked	68	Dirty guttering	66
3.	Polluted surface water within 10 m	67	No filter medium	62
4.	Ponding on plinth	66	Tap leakage	57

It has been shown that dug wells are far more vulnerable to contamination than tubewells since there are more potential points of risk and a larger area to protect. Although dug wells are the most high risk of the facilities investigated they will none the less remain the principal source (>90 per cent) of rural water for many less developed countries in the medium term and are therefore a principal focus for remedial action strategies.

Conversion of dug wells should be the most economic strategy for improving water quality but clearly requires more careful supervision and reinspection to ensure that significant risk reduction results from the remedial action interventions. At the present time a concrete slab is placed over the dug well, a handpump attached to the slab and a suction pipe hung into the well. It is therefore not surprising that water quality improvement is marginal. We therefore recommend the following additional points of improvement:

1. The inside wall of the well should be rendered with mortar for 3 m depth below the surface of the ground, or be replaced with concrete rings.
2. The walls above ground (parapet) should be replaced with a ferrocement plinth, covering the well to an area of at least 4 m^2 and correctly sloped to lead excess water to drainage.
3. The drainage pipe should be at least 10 m in length.

4. The plinth should be fitted with a mild steel sanitary cover with a locked inspection manhole.
5. The well should be fitted with a handpump, delivery pipe and well screen bedded securely in coarse sand or fine gravel.

In Java the total costs of the conversion (1)–(5) was about $US 125 at 1988 prices. This compared with approximately $US 250 for a complete shallow tubewell installation and up to $US 650 for a new deep handpumped tubewell. The tubewells were therefore targeted at larger communities and dug well conversions for smaller neighbourhoods of 5–10 families or even individual families.

8.3 Zambian pilot project

The approach adopted for water supply improvement in Mongu district was based less on remedial action of organized supplies following risk assessment, and more on fundamental development of primitive, unprotected traditional sources. The sanitary inspection forms for risk assessment presented in Chapter 5 were first used in Indonesia and Peru pre-1989. Since those pilot trials their use has been extended to Thailand and also to other countries in Latin America. They were not used by the DWA in the Mongu district of Zambia. The CEP/WASHE and health assistants in Mongu provided information for improvement and rehabilitation of traditional point sources using other criteria. They focused their improvement strategy on uprating well points, shallow wells with windlasses and handpumped tubewells using a 60-point scoring system devised by the DWA and approved by the provincial WASHE committee. The selection criteria included:

1. adequate population size, greater than 30 households or >180 people;
2. accessibility of the source: not more than 15 minutes walking time from the furthest household;
3. reliability of the source: year-round continuity of the supply;
4. quality of the source: bacteriological and visual turbidity;
5. indicators of community participation.

Thus it can be seen that most of thee critical indicators of quality of service, except relative sanitary risk, were used in selecting sources for improvement. None the less, sanitary inspection was carried out although in the absence of comparative risk data the selection strategy depended more heavily on the quality data.

The dissemination of surveillance results stimulated varying degrees of response: 36 sources were reported to have been improved on the initiative of health assistants, and all but one of these were unprotected sources, to which protection was increased. Few attempts seem to have been made to resample and see whether quality improved.

Results of sampling passed on to the DWA were used by the district teams to give highest priority in maintenance to those wells showing highest levels of contamination. However, as the wells sampled by health assistants constituted only about half of the protected sources in the district, separate multiple sampling was done by the DWA for all shallow wells and some handpumps. The results of this sampling constituted about half of the samples analysed.

Since handpumps were found to exhibit very low contamination levels, most sampling was concentrated on traditional sources and shallow protected wells. All the latter identified by sanitary and selection criteria as needing rehabilitation were repaired during the period of the project. Sampling of some 46 of these over 18 months showed that 20 had no faecal coliform at any time, 21 improved in quality, 2 got slightly worse, and 3 remained grade B (<10 faecal coliforms/100 ml) of contamination throughout.

As a result of the microbiological analysis of water and the rehabilitation programme, a simple method of well chlorination was devised using locally available laundry bleach. When a well had been bailed down so that only 0.5 m of water remained, one bottle (800 ml) of bleach produced on average 50 mg/l free chlorine after 10 minutes, and 25 mg/l after 2 hours. Bottles which had been on the shelf for more than 6 months gave 12–17 mg/l free chlorine. Chlorinated water was bailed out after 2 hours, and used to wash down the sides of the well and the surrounding slab. In all cases wells were found to be coliform free after this treatment. Chlorination was not carried out until after remedial works to remove the source of contamination.

8.4 Peruvian pilot project

It was indicated in section 7.3.8 that the Peruvian diagnostic survey revealed that the supplies incorporating treatment were generally high risk. This survey demonstrated that the majority of system components had fundamental operational and design problems. These findings led to the development of a strategy for rehabilitation and improvement of treatment plants. The 1986 survey included one urban and 17 small treatment plants in 2 departments, Huancavelica and Junin. Two of the plants had rapid sand filters which were not working. The results for all 18 filtration plants are summarized in Table 8.2 and reveal that all slow sand filters and disinfection units have major deficiencies and operating problems. The reasons for this have been examined in detail and a clear pattern of common problems has emerged. They can be grouped in two main categories, administrative and technical.

Systems of the PNAPR are constructed jointly with the community which provides the work-force. The community is required to establish an administrative committee (JAAP) responsible for operation, maintenance and recovery of operating costs. The administration and populations served by the schemes in the study area are summarized in Table 8.3. This shows that the majority of rural schemes fall into the JAAP administrative category and herein lies a major

Table 8.2 Rural water treatment systems in the central highlands and high jungle of Peru (Lloyd *et al.* 1986b)

		Systems presenting major deficiencies and problems	
System component	Sites surveyed	No.	%
Abstraction	18	16	89
Settlers	18	11	61
Slow sand filters	16	16	100
Disinfection	18	18	100

Note: The Ministry of Health reports a total of 28 rural treatment systems in the region.

problem in supplying a safe, continuous supply of water. The JAAP is not trained, neither are there any incentives to provide a professional service. Until recently the committees have almost never received professional supervisory support from the Ministry of Health after supplies have been commissioned. Therefore a pilot project to establish a training and a supervisory infrastructure through DITESA was set up.

Although some of the technical problems of treatment relate to faulty construction, the more fundamental problems are those associated with the raw water source quality and flows for which the standard designs are inadequate. With grossly contaminated source water it is essential to build a series of barriers to prevent contamination entering into distribution. This process of protection and treatment begins with source water selection and abstraction. It is clear from Ministry of Health records that although a chemical analysis is carried out on intended source waters the more important turbidimetric and sanitary analysis is neglected and no adequate design provision had been made to cope with high turbidity and faecal contamination.

8.5 Technology example – Peru

Of the three pilot projects it was the Peruvian which went all the way from surveillance to control and even to the rehabilitation of default systems. Peru is, therefore, used as an example to demonstrate alternative technological approaches to improving drinking water quality. The main processes used in Peru are filtration and disinfection.

The problems of performance of slow sand filters cannot be considered in isolation from the other components of the system. In order to provide a better understanding of the whole, the technical problems identified are summarized in Table 8.4 and described here in sequence from source water abstraction to terminal disinfection. A typical treatment plant is shown schematically in Fig. 8.2.

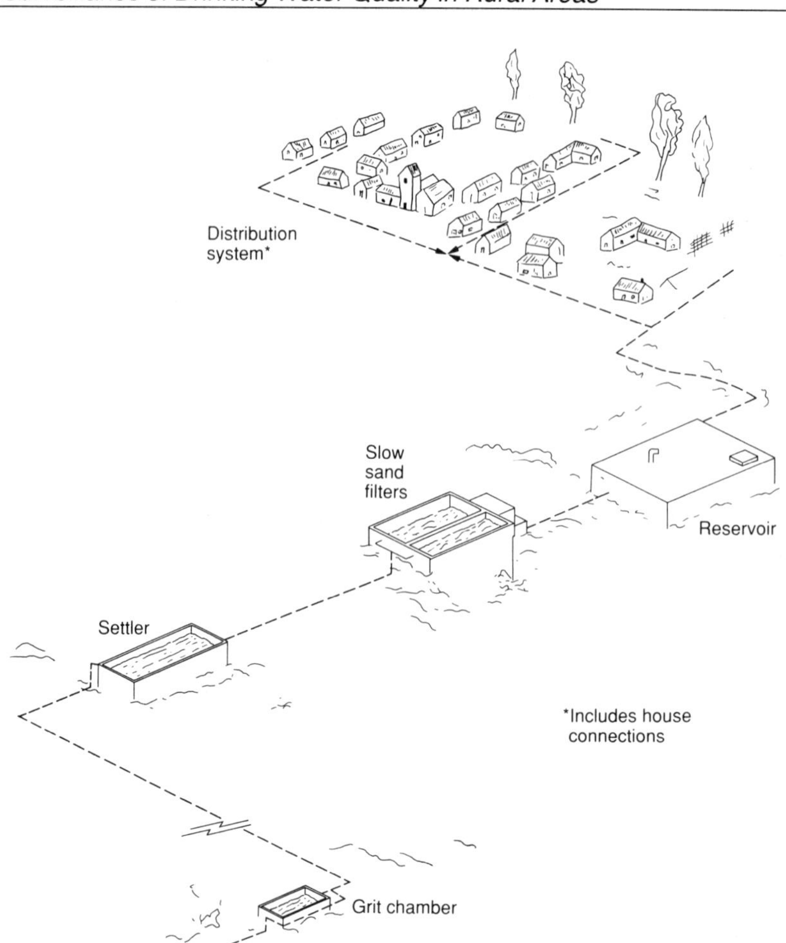

FIG. 8.2 Layout of a typical slow sand filtration plant constructed in Peru since the 1970s.

8.5.1 *Abstraction points intake*

Water treatment processes are designed to work at a controlled flow rate and the standard design of the systems in Peru rely on flow control at the point of abstraction. This includes a 90° V-notch weir which is inappropriate for measuring flows of less than 70 m^3/day (0.8 l/sec). The diagnostic revealed the following:

1. In 16 out of 17 rural plants the V-notch was not installed and should in any case be a 45° angle for this type of flow.

Table 8.3 Summary of treatment systems in the Peruvian water surveillance programme (1986a)

Community	Administrative authority	Total population	Population served	House connections
Huancavelica	SENAPA*	17,452	11,892	2,500
Yauli	JAAP†	1,868	1,230	300
Palian	JAAP	1,334	693	228
Cocharcas	JAAP	1,195	355	100
San Martín de P	JAAP	Periurban	229	50
Tres de Diciembre	JAAP	1,002	248	50
San Agustín de Cajas	JAAP	4,246	1,265	230
Chaquicocha	JAAP	667	479	87
San José de Quero	JAAP	1,307	644	137
Huayao	JAAP	655	480	80
Churcampa	Council	1,859	715	140
Hualhuas	JAAP	1,751	1,375	233
Saños Grande	JAAP	5,000	1,750	350
El Mantaro	JAAP	3,016	2,280	420
Julcan	JAAP	2,126	1,167	200
Sacsamarca	JAAP	2,205	845	222
Tarmatambo	JAAP	2,042	1,370	234
Pichinaki	Council	1,890	1,364	600

* SENAPA: National Water and Sewerage Authority – Urban authority.
† JAAP: Community drinking water administrative committee – 'rural' authority.

2. In 17 out of 17 plants the intake was poorly constructed or unprotected, either lacking a screen, or too coarse or broken.
3. In 15 out of 17 systems there was no weir in the source water canal to ensure that there was a continuous minimum head; consequently the majority of systems suffer regular discontinuity of flow. This is often aggravated by neighbouring communities cutting the flow to the plant in favour of their perceived irrigation requirements. It is not well understood that the proportion of water required for potable supplies is insignificant compared to that required for agriculture.

8.5.2 Settlers/sedimenters

These are generally subjected to short circuiting through inadequate capacity and lack of baffling. Spot checks using the transit time of inlet salt dosing revealed minimum retention times of 10–20 minutes and in 16 out of 18 systems there was no form of baffling.

Settler efficiency is further reduced by lack of routine cleaning. Their efficiency in reducing turbidity is rarely more than 30 per cent and consequently they fail to protect the slow sand filters which quickly become blocked. In

Table 8.4 Summary diagnostic of technical problems in Peruvian pilot region water supply systems (Lloyd and Pardon 1988)

Community	Abstraction point	Sedimenters	Sand filters	Reservoir	Distribution	Quality in supply		
						Chlorine residual (mg/l)	Turbidity (TU)	Bacterial (Grade)
Huancavelica (urban)		+ (coagulation)	++	+		0.3	<5	A
Yauli		+ (coagulation)	+		++	00	15–32	D
Palian	+++	+	++++	+		00	15–>30	D
Cocharcas	+++	++	++++			00	<5–>25	D
San Martín de Porras	++	++ (coagulation)	++	++		00	10–20	D
Tres de Diciembre	++	+	+	+		00	<5	B
San Agustín de Cajas	+++	++	+++	+	++	00	15–50	D
Chaquicoc	++	++	++		+	00	<5	C
San José de Quero	+	++	++		+	00	<5	B
Huayao	+++	+++	++	+	+	00	<5	C
Churcampa	++	+++	++		+	00	<5–9	C
Hualhuas	++++	+++	++++	+	++	00	15–>50	D
Saños Grande	++++	+++	++++	++		00	18–500	D
El Mantaro		+	+++	++		00	10–60	C
Julcan	++		+	+	+	00	<5	C
Sacsamarca	+++	++	+++	+	++	00	6–>50	D
Tarmatambo	+++	++	+	++	+	00	<5	C
Pichinaki	+	+	+			00	<5	C

+ = No. of problems identified at each stage.
Bacterial grade: A = 0, B = 1–10, C = 11–50, D = >50 faecal coliforms/100 ml.

evaluating raw water quality prior to construction of plants no turbidity or suspended solids analyses have been made during rainy periods in order to size the sedimenters correctly.

8.5.3 *Slow sand filters*

The two fundamental problems presented by slow sand filtration in Peru and many other countries are: (a) their inability to cope with *high turbidities* (normally in rainy periods), and (b) *flow variation*. If turbidity consistently exceeds 15 TU filters may block in several days unless protected by pretreatment. In addition unwashed and incorrect grading of filter sand leads to short filter runs.

In the Peruvian diagnostic survey it was repeatedly observed that the absence of flow control at the point of abstraction caused filters to be operated at an unstable rate, intermittently. It is well known that this results in sub-optimal performance, for example in terms of low bacteriological removal efficiency. More than half the plants had marginal or no effect in reducing turbidity and contamination. Table 8.4 demonstrates that a majority of treatment systems produced grade D water, routinely passing greater than 50 faecal coliforms/100 ml into supply. Furthermore, almost all of the systems with slow sand filters regularly supplied faecally contaminated water; only 3 out of 16 slow sand filters significantly reduced faecal coliform counts to less than 10/100 ml (grades A and B).

Table 8.5 highlights the serious neglect of basic operation and maintenance in slow sand filters:

1. in 16 out of 16 plants sand bed depth was below the recommended 60 cm minimum and two filters had no sand at all;
2. in 16 out of 16 plants there was no installation of designated area for sand cleaning and storage.
3. None of the filters had any means of measuring the amount of water leaving the filter.

8.5.4 *Disinfection*

A diffusion hypochlorinator is recommended by the PNAPR which is designed for installation in a constant 1 l/sec flow. This was occasionally found installed on a retaining wire under the inspection cover inside reservoirs. In 7 out of 16 systems the hypochlorinator was found hanging in the reservoir at points of convenient access but in none of these was it located in a flow of water.

On no occasion was a measurable chlorine residual detected at the outlet of any rural reservoir or in distribution in any of the rural systems. There was an urgent need to optimize the design and reassess the point of installation of this device. This was the subject of a separate investigation and development project sponsored by the ODA of the British Government.

Table 8.5 Operational problems and remedial action proposals required for Peruvian rural slow sand filters (Lloyd *et al.* 1986b)

Community	Depth of sand bed* (cm)	Volume of sand reserve (m^3)	Distance to source of sand (km)	Actions required
Palian	20	00	25	P
Cocharcas	00	00	5	P
San Martín de Porras	30	00	10	R
Tres de Diciembre	25	200	0.12	R, C, I
San Agustín de Cajas	55	00	3	P
Chaquicocha	10	00	10	R, C, I
San José de Quero	40	00	12	R, C, I
Huayao	50	00	1	R, C
Churcampa	00	00	?	P
Hualhuas	35	00	18	P
Saños Grande	20	00	25	P
El Mantaro	10	00	?	P
Julcan	40	00	15	C, I
Sacsamarca	20	00	50	P
Tarmatambo	35	4	55	R, C, I
Pichinaki	40	00	?	I

* Minimum sand depth recommended is 60 cm.
P = complete rehabilitation; R = complete sand bed replacement; C = cleaning; I = increase depth of sand.

8.6 Rehabilitation project – Peru

It was concluded from the diagnostic survey that the communities at greatest risk from water-borne disease were those with no option other than surface-water sources since the existing treatment systems had failed to reduce critical contaminants to safe levels. As a result of the survey the Ministry of Health imposed a moratorium on the construction of new slow sand filter installations until pilot demonstration projects could demonstrate improved performance.

It was clear that in those supplies where the raw water turbidity overloaded the filters rehabilitation and new construction, incorporating prefiltration, were required. The systems classed as grade D bacteriologically, required priority attention to improve both the quality and sometimes also continuity of the supply.

The first pilot demonstration project to incorporate gravel prefiltration in Peru was a complete new treatment and supply constructed for the community of Azpitia (Lloyd *et al.* 1986c; Lloyd and Pardon 1988). Prior to this only two slow sand filter plants had been improved, as shown in Table 8.6. These supplies, at Carhua and San Buenaventura, did not require gravel prefilters, since their raw water turbidities rarely registered more than 20 TU even in periods of heavy rainfall. These two schemes were improved by sand replacement using appropriately graded filter sand and, in the case of San Buenaventura, by using synthetic matting to enhance performance and protect auxiliary 'Potapak' package plant slow sand filters.

The Azpitia treatment plant has a small settler integrated with a three-stage *vertical downflow* gravel prefilter in one structure and four parallel 'Potapak' slow sand filters. The scheme was completed in early 1985 and performance evaluation was carried out prior to the commencement of the first rehabilitation project of an existing slow sand filtration plant in the pilot project region at Cocharcas.

The Cocharcas rehabilitation project was completed in 1986 and incorporated the first three-stage *horizontal* flow gravel prefilters constructed in Peru as shown in Figs 8.3 and 8.4. The pilot rehabilitation project for Cocharcas (Pardon 1989) incorporated the following components:

1. Administrative, joint project planning with the community water administration committee and public meetings to ensure majority support and community participation in the proposed intervention:
2. Reconstruction of abstraction box: to ensure minimum flow, flow control and source protection:
3. Construction of three-stage horizontal flow gravel prefilters: to protect slow sand filters from overload by turbidity and suspended solids, and provide multiple barrier treatment:
4. Rehabilitation of slow sand filters: to provide resanded beds, graded gravel under drainage and flow control to optimize performance:
5. Terminal hypochlorite disinfection: to polish the filtered water:
6. Water committee training: to ensure proper operation and maintenance post-commissioning:
7. Operations evaluation: to establish sustained improvement in the water supply service.

All of the above were necessary to ensure that the supplies were improved from high risk to low risk.

We are confident that the filtration technologies can perform satisfactorily; the more difficult task ahead is to develop a level of administrative infrastructure which will ensure an adequate level of operation and maintenance of these systems.

An incremental programme of remedial action and rehabilitation has been formulated as part of the national strategy for preservation of rural water supply services in Peru. Table 8.6 emphasizes that in the first stages this was limited to the improvement and rehabilitation of rural treatment. However, in some localities it was demonstrated that the source water for treatment was actually a spring. At Hualhuas, for example, although rehabilitation was required it was shown to be more cost-effective to protect and pipe the spring source to distribution rather than incur the costs of rehabilitation of treatment. It is generally worth piping reliable spring sources up to 10 km rather than involve

Table 8.6 Improvement and rehabilitation strategies in rural water treatment plants in Peru* (Pardon 1989)

Year of intervention	Location	Department	Type of intervention	Prefiltration	Enhanced slow sand filter	Training of community	Follow-up evaluation
1984	Carhua	Lima	I		•		•
1984	San Buenaventura	Lima	I		•		•
1985	Azpitia	Lima	N	•	•	•	•
1985	Iscozacin	C. de Pasco	I		•	•	•
1986	Espachín	Lima	N	•	•		
1986	Cocharcas	Junin	R	•	•	•	•
1987	La Cuesta	L. Libertad	R	•	•	•	•
1987	Compin	L. Libertad	R	•	•	•	•
1988	Palian	Junin	R, C	•			
1988	Viccos	Junin	N, C	•	•		
1988	Collambay	L. Libertad	N, C	•	•		
1988	Cayanchal	L. Libertad	N, C	•	•		
1988	Simbal	L. Libertad	N, C	•	•		
1988	Hualhuas	Junin	R, PBP			•	•

* Information up to July 1988/Rural communities, <2000 inhabitants.
I = improved slow sand filters; N = new scheme; PBP = protect spring and bypass treatment; R = rehabilitation of existing scheme; C = in construction.

the community in the complexities of treatment plant operation and maintenance.

Elsewhere in the Peruvian pilot region coverage in communities is being increased from less than 50 per cent to greater than 90 per cent by extending the distribution systems and numbers of connections as an integral part of the improvement programme. It is clear that increasing coverage is still the essential prerequisite before the other surveillance indicators can be seriously attended to.

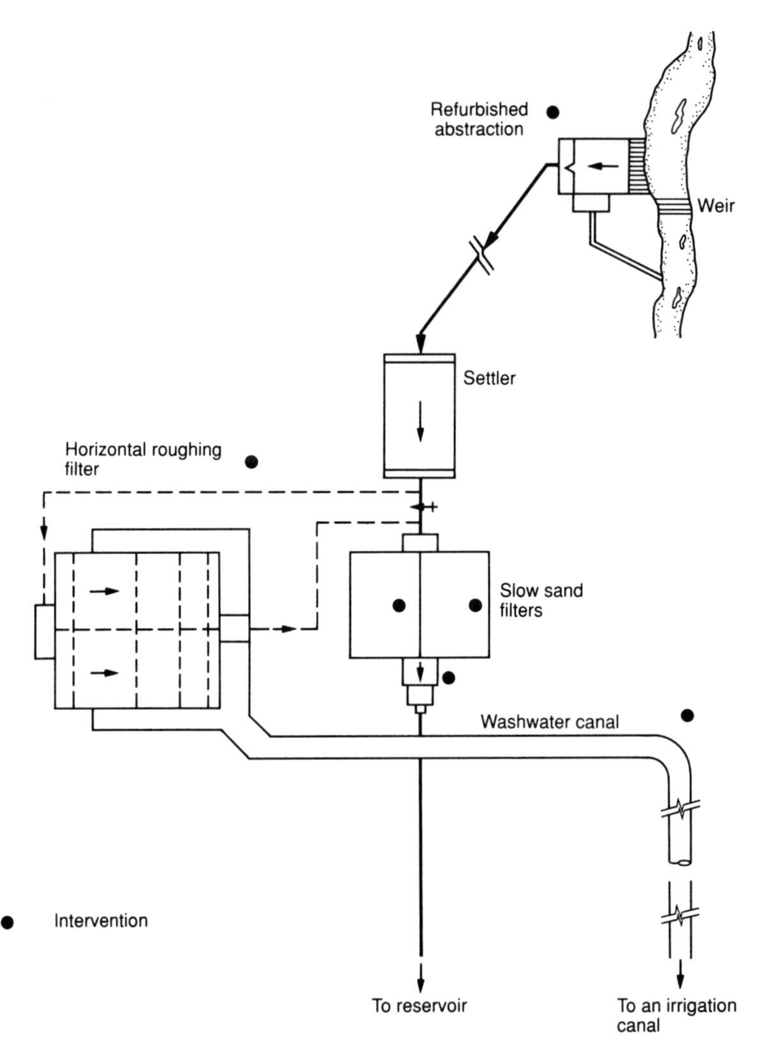

FIG. 8.3 Schematic drawing of the rehabilitated water treatment plant of Cocharcas.

FIG. 8.4 Community work at the Cocharcas horizontal prefilters.

CHAPTER 9

WHO Strategy for Technical Cooperation

9.1 Programmes and linkages

Since the early planning stages of the IDWSSD it was obvious to the international organizations concerned that the attainment of the target, i.e. safe water supply and sanitation for all by 1990, required new approaches both in national strategies and in external support (WHO 1981).

The IDWSSD also represents an essential first stage of the global programme of health for all by the year 2000. Increasing the quantity and quality of drinking water would help to reduce the incidence of many diseases among the people most at risk. Improving sanitation and personal hygiene could greatly enhance the health impact of investments in water supplies. It is these health aspects of water and sanitation which concern WHO the most, and which provide the link with its primary health care goals and programmes.

Programmes for health and water quality surveillance were promoted from the early years of the IDWSSD. Planning of relevant projects was based on considerations that water supply improvements should not be introduced without sanitation and health education; piped water systems where water treatment produces a product of a lower standard than traditional sources are useless; and water development schemes (whether for irrigation, dams or reservoirs) that increase the risks of tropical diseases are worse than useless. This underlines the increased importance, for health and water quality, of surveillance programmes, which do not necessarily imply more investment in expensive laboratories and training facilities for highly qualified laboratory technicians. The first priority is to provide for areas where services are known to be deficient and health risks high. Yet monitoring, surveillance and remedial action for disease prevention must be applied to all IDWSSD programmes.

But what has become of these idealistic intentions with which the IDWSSD was set into motion? And what has been done about the hygiene and safety of water supplies? The WHO is following a strategy for technical cooperation with Member States on the control of environmental hazards which outlines an organization-wide strategy in the area of drinking water quality with the following three steps (WHO 1987a).

9.1.1 The situation during the Decade

Although the number of people being adequately supplied with drinking water drastically increased during the first half of the IDWSSD (see Fig. 9.1), there was much less progress on the safeguarding of the quality of the water provided (WHO 1988d). Communicable water-related diseases, with diarrhoeas in first place, are still the most widespread health problem, particularly in the under-served rural areas of developing countries. Appropriate measures to protect drinking water quality, not only from microbiological contamination, but also from chemicals, are still needed in many countries. Lack of human and financial resources severely hampers the public health and public works authorities in discharging their responsibilities with regard to drinking water quality surveillance and control. National drinking water quality standards, where they exist, are often not supported by the necessary laboratory services to monitor their compliance or to stimulate improvements in the safety of water supplies.

Even in urban areas where major improvements have been made in the production and distribution of good quality water, contamination of water supplies occurs within the household itself during handling and storage.

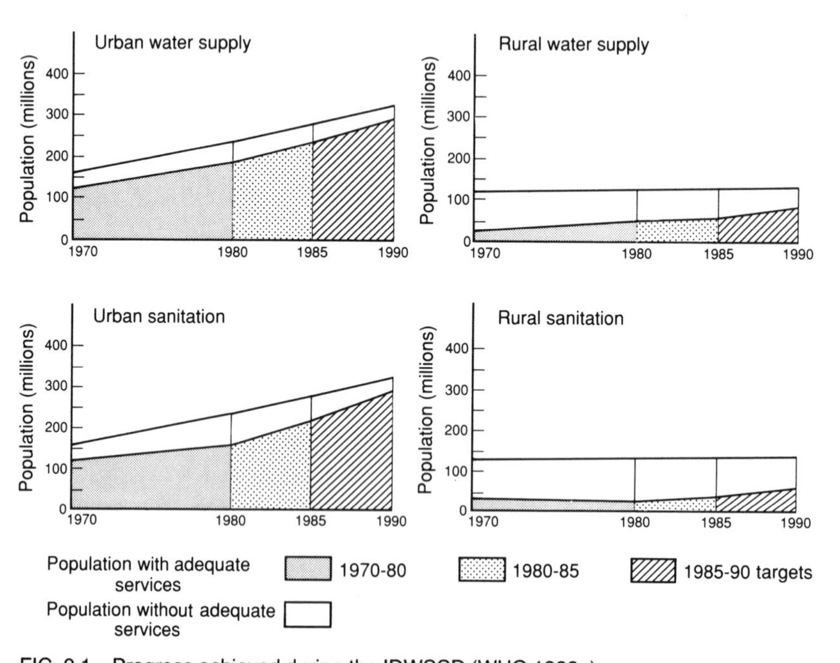

FIG. 9.1 Progress achieved during the IDWSSD (WHO 1988a).

9.1.2 *Strategies for the 1990s*

There should be a dual target regarding drinking water quality. Firstly, where the existing infrastructures and resources permit, national drinking water quality standards should be formulated to support countrywide improvement of drinking water quality. The implementation of such standards must be accompanied by practical and feasible surveillance and with the provisions and means to take remedial action when required (WHO 1988c).

Secondly, in rural areas and small communities where standards as such have little meaning, action plans need to be developed and implemented to improve the protection of water supplies from bacteriological contamination (Helmer 1989). This will require regionally or locally based systems linked to primary health care for raising public awareness of the problem and possible solutions, and for implementing minimal water quality surveillance and providing remedial measures with reliance on appropriate technology and community participation. National/local programmes are needed to aim at reducing the contamination of drinking water during storage and within the dwellings themselves. This will require health education, guidelines, surveillance, etc.

9.1.3 Programme goals for technical cooperation

WHOs technical cooperation with its Member States is based on the application at country level of the recently issued WHO Guidelines for Drinking Water Quality (WHO 1984/85). This cooperation concentrates on:

1. Promotion and translation of the WHO Guidelines and use at national and local level by public health officers and water supply agencies:
2. Conducting regional and national workshops, followed by consultant/staff missions, to support the development and application of national standards for drinking water quality:
3. Support of national demonstration projects on managerial and technical approaches for instituting drinking water quality protection in rural areas including the training of staff:
4. Strengthening the institutional and manpower capabilities of health agencies to fulfil their function as surveillance agencies:
5. Providing guidance to local and district agencies and, where applicable, by primary health care approach to prevent secondary water contamination during household use.

The magnitude of these tasks, particularly due to the size of the rural populations to be reached, called for a change in the usual methods of delivering technical support. Close cooperation was established early in the IDWSSD with other international organizations and external support agencies. Particularly in drinking water quality, previous linkages with the UNEP concerning the quality of freshwater resources were expanded leading to substantial support from this agency (Helmer and Ozolins 1987).

Since its creation by the UN Conference on the Human Environment in 1972, UNEP has helped tackle problems related to irrigation, water transfer, drinking water supply and water quality management. Practical outputs have been the formulation of guidelines, the publication of research results and numerous training courses and seminars. In 1986, UNEP launched the programme for the Environmentally Sound Management of Inland Waters (EMINWA). This programme will help develop inland water systems, train local staff, provide guidelines on water quality management, make regular global assessments of the state of inland waters, and increase public awareness of the need for environmentally sound water source development. It is under the EMINWA programme that UNEP contributes to the IDWSSD and, since 1984, supports the three-country pilot project which is the main subject of the present publication (WHO 1989a). The Zambian project was also part of the Zambezi Action Plan (see Fig. 9.2).

Another joint venture of UNEP and WHO, which is intimately linked to drinking water, is the Global Freshwater Quality Monitoring Project (GEMS/WATER) under which the concentration of chemicals and bacterial indicators are routinely monitored at more than 300 sites worldwide in rivers, lakes, reservoirs and groundwater aquifers (see Fig. 9.3). Most of these resources are used for public supply of drinking water. Technical cooperation with Member States includes the strengthening of water quality monitoring, the improvement of data quality and reliability through analytical quality control, and the assessment of long-term trends of water pollution by hazardous substances. All monitoring data are globally processed at the WHO Collaborating Centre for Surface and Groundwater Quality at the Canada Centre for Inland Waters in Burlington, Ontario (Barabas 1986). Analytical quality control check samples are distributed from USEPAs (United States Environmental Protection Agency) Environmental Monitoring and Support Laboratory in Cincinnati to all participating laboratories. USEPA also provides back-stopping in analytical procedures (Winter 1985). It is primarily the methodology developed under this project which is of immediate use to laboratories charged with the surveillance of drinking water quality. Thus, the *GEMS/WATER Operational Guide* and the recommendations for within-laboratory quality control provide advice on essential surveillance methods for sampling, analysis and data processing (WHO 1987b).

9.2 Delivery of technical cooperation

Technical support for developing countries, in general, is provided by WHO on request from Member States. Ideally, such requests should emerge as an integral part of a comprehensive national action plan for the water supply and sanitation sector and/or a national programme for primary health care. The operational responsibility for the delivery of technical cooperation rests with the six WHO Regional Offices, each of which have an environmental health unit. In addition, several regional/sub-regional centres provide technical service and services in the area of the drinking water supply and water quality control (WHO 1987a).

In the African Region, there are three sub-regional WHO intercountry health development teams – in Bamako, Bujumbura and Harare – each with expertise in sanitary engineering. In the region of the Americas, the Pan-American Centre for Sanitary Engineering and Environmental Sciences (CEPIS) in Lima

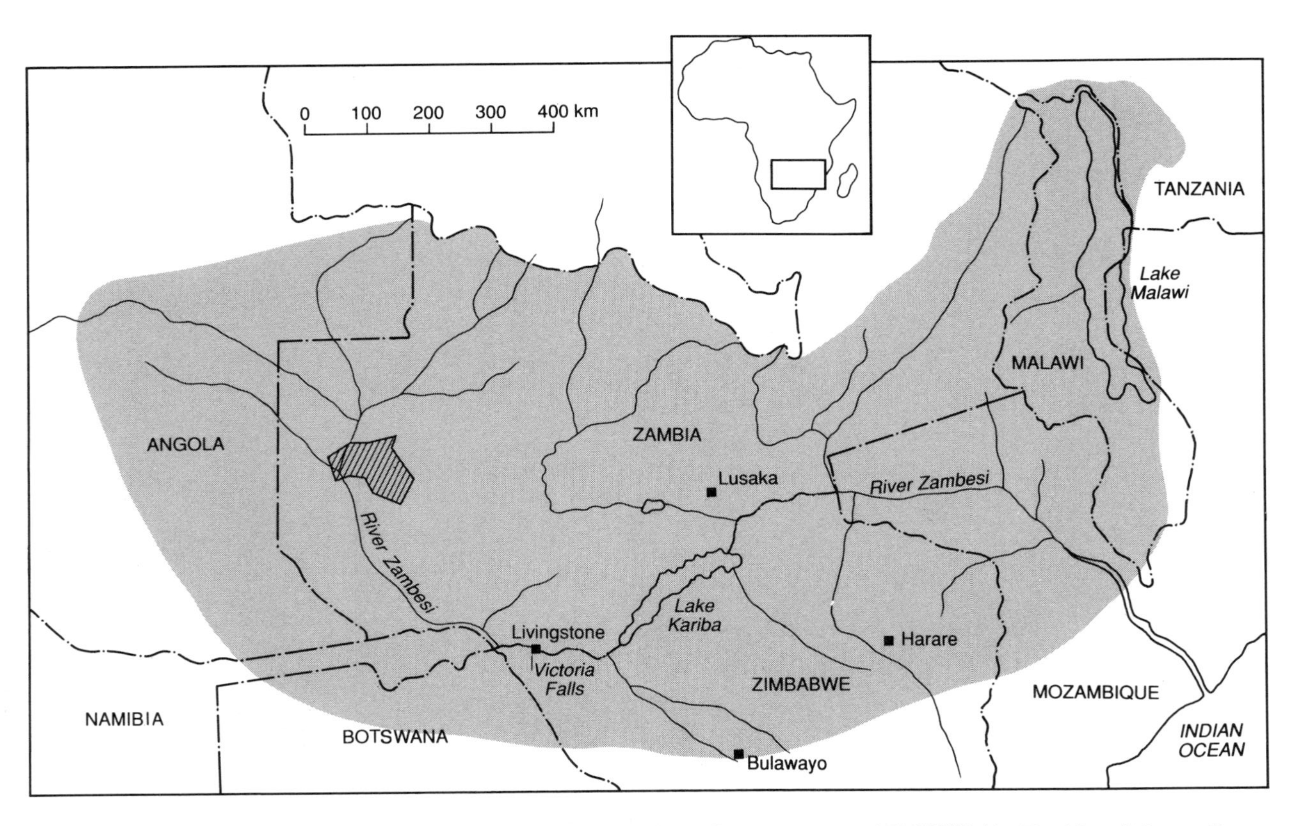

FIG. 9.2 The Zambezi basin. The Zambezi Action Plan area which is a first target area of EMINWA (the Zambian pilot surveillance project (hatched area) is located in the river's flood plain). Source: UNEP, undated.

FIG. 9.3 GEMS/WATER global distribution of monitoring stations (groundwater, rivers and lakes), September 1985 (WHO 1986).

supports regional projects; it was already instrumental in getting the Peruvian pilot project established. The Regional Centre for Promotion of Environmental Planning and Applied Studies (PEPAS) in Kuala Lumpur, Malaysia, provides similar services to the WHO Western Pacific Region. Also, the WHO Eastern Mediterranean Region established such a technical arm, the Regional Centre for Environmental Health Activities (CEHA) in Amman, Jordan, which conducted an impressive series of national seminars to promote understanding and application of the WHO Guidelines (see Table 9.1). Depending on the size of the WHO programme, there are WHO country engineers posted on site to assist in the implementation of IDWSSD-related projects in all regions.

Table 9.1 National seminars and consultancies conducted by the WHO Region Office for the Eastern Mediterranean in support of the WHO Guidelines for Drinking Water Quality

Venue	Month	Year
Amman, Jordan	September	1985
Khartoum, Sudan	March	1986
Kabul, Afghanistan	July	1986
Baghdad, Iraq	July	1986
Amman, Jordan	December	1986
Mogadishu, Somalia	March	1987
Khartoum, Sudan	March	1987
Alexandria, Egypt	November	1987
Damascus, Syria	November/December	1987
Islamabad, Pakistan	December	1987
Cairo, Egypt	October	1988
Rabat, Morocco	November	1988
Hammamat, Tunisia	November	1988
Muscat, Oman	December	1988
Islamabad, Pakistan	February	1989

Much of the technical documentation and preparation of methodology guidance on drinking water quality issues, including the WHO Guidelines, has been prepared at WHO Headquarters, often in response to specific regional needs. Since the publication of the WHO Guidelines, a large variety of programme activities have been launched by WHO and its Regional Offices with the aim of promoting the application of the WHO Guidelines. The generally recognized need for support of developing countries on this subject has led to a rapid increase in international (UNEP, UNDP) as well as bilateral (e.g. DANIDA, GTZ, NORAD, ODA) donor funding of projects during the second half of the IDWSSD. Scientific and technical advice became much in demand from WHO which is the specialized agency most concerned about drinking water quality and associated human health effects. Continuous and reliable back-up by a competent and resourceful institution became a mandatory requirement if the professional level and scope of response to country needs were to be maintained by the organization during the end-IDWSSD and post-IDWSSD period.

The close cooperation maintained between DelAgua and CEPIS throughout the Peruvian pilot project was put on more formal grounds in 1988 with the designation of the Robens Institute, and particularly its environmental health unit, at the University of Surrey, UK, as the 'WHO Collaborating Centre for Protection of Drinking Water Quality and Human Health'. It was agreed that the Collaborating Centre participates in and actively supports the work of the Organization in the area of drinking water quality control, particularly in developing countries.

The present publication on the three pilot projects is the first prominent outcome after the centre's designation, although cooperation with WHO dates back to the development of the first prototypes of the DelAgua water testing kits which were then used in the pilot areas and related training courses. Many of the examples of technical cooperation described in the following section are drawn from this collaboration programme.

9.3 Programme elements of the strategy

Many of the activities undertaken by the WHO to promote effective drinking water quality control in rural areas revolve around the pilot projects described in the present publication. Thus, it is evident that the methods and techniques developed for and applied in these projects provided the basis not only for other similar country initiatives but also for a host of training courses, seminars, on-site education, etc. Together with the newly designated WHO Collaborating Centre for Protection of Drinking Water Quality and Human Health an array of activities were brought together in a coherent programme. Some of the major elements are as follows.

9.3.1 Information exchange

A wealth of data, reports and studies accumulates continuously from national and regional projects, action plans, research undertakings, etc. thus allowing for comparative evaluation of experience which is put together and used in other countries with similar water resources or similar socio-economic infrastructure. By way of periodic publications, translation into other major languages, dissemination of reports and manuals, etc. a large audience is reached through the network of WHO country contacts, through professional associations, and through national or bilateral development organizations. Relevant drinking water issues were regularly dealt with in the *Water Quality Bulletin* issued quarterly by the WHO Collaborating Centre for Surface and Groundwater Quality, Canada, in English and French. Technology news in Spanish is periodically disseminated by CEPIS, Peru, by way of a newsletter entitled *Hojas de Divulgación Técnica*. Also, the WHO Regional Offices compile standard model designs for rural water supplies (WHO 1976b; WHO/EURO 1986) and issued a review of available drinking water chlorination methods (WRC 1989).

The WHO Guidelines, volume 3, on drinking water quality control in small community supplies was translated not only in major languages such as English, French and Spanish, but has also been issued in Indonesian and will appear in Arabic. Experience in our pilot projects has shown that district health inspectors and local water supply agents can often not be reached if local languages are not used.

9.3.2 *Harmonization of methodology*

Field tested approaches to all steps in the surveillance and control process have emerged from the pilot areas and other country projects. As an immediate outcome, improved standard protocols are now available for the planning of surveillance schemes, more practical forms for sanitary inspection are at hand, and concrete advice for the improvement and rehabilitation of faulty water supply installations can be given. Examples and details of these methods were given in Chapters 6 and 8.

The most important output of these efforts to streamline and consolidate current approaches and methods will be the revision of the WHO Guidelines, volume 3 on drinking water quality control in small community supplies (WHO 1986). Thus practice-proven guidance will be brought together at the end of the IDWSSD to lead the way in drinking water quality control for the post-IDWSSD period where quality issues are expected to receive even more attention.

9.3.3 *Appropriate technology*

Since the early years of the IDWSSD the techniques and equipment for water quality surveillance underwent in-depth re-evaluation with the aim to make them simpler, cheaper and more appropriate to the requirements of small communities and the level of technology and skills available to health and water agencies in rural areas. Two avenues were explored: the basic district laboratory and, as an alternative, the field kit for on-site water testing.

The design and needs for stationary water laboratories at the basic, intermediate and central agency level were outlined in a WHO guidance document entitled: *Establishing and Equipping Water Laboratories in Developing Countries*, which is of great help in selecting the necessary equipment, purchasing consumables and planning the outputs in terms of samples processed and analytical results obtained. Costs of installation and operation, as well as staff and skill requirements, are indicated on the basis of experience with country projects (WHO 1986).

As concerns field kits for bacteriological and chemical analysis, several independent research groups – in countries such as China, India and the UK –

scrutinized the specifications and needs for simple field tests (DelAgua 1988). The results are today a choice of field kits on the market which not only contributed to an affordable price for such kits but also to an optimization of instruments, equipment and power supply alternatives. In combination with a suitable vehicle, e.g. motor cycle, great mobility has been reached and even remote rural countries can receive a minimum of surveillance services. One fundamental aim of the WHO strategy has thus become feasible: to undertake sanitary inspection combined with sampling and basic bacteriological and physico-chemical testing of all supplies in a given health district at reasonable time intervals, so that at least two visits per year could be made.

Appropriate technology development does not stop, however, at the surveillance phase but equally has to cover the process and intervention needed to correct defaults and to eliminate contamination of water supplies (Dahi 1989). Therefore simple treatment for sediment removal has been designed and put into use, such as horizontal flow filters and other types of roughing filters to protect slow sand filters and secure their optimal functioning. Examples from the Blue Nile Health Project in Sudan and from the Andes in Peru confirm the suitability and validity of their conceptual design. In many situations, however, disinfection by adding chlorine products in one form or another, are the ultimate resort to guarantee hygienic safety of the water supply. Simple mechanisms to supply the needed chlorine dosage were invented and put to use. Much progress has been achieved in developing electrolytic chlorinators and a similar process is to facilitate the *in situ* generation of the disinfectant independent of unreliable supplies. Various trials are supported by WHO and its Collaborating Centre which submits various disinfection methods to comparative testing and evaluation. Pilot projects are also used for long-term trials under field conditions.

9.3.4 Training of health and water agency staff

Pursuant to the dual philosophy of drinking water quality surveillance and control, the water supply agency as well as the health authority responsible for the area should have their respective staff trained for the tasks involved. The crucial importance of the human factor was recognized early in the IDWSSD and has since then evolved in a multitude of training activities ranging from the local to the global level.

An example was set in the WHO Eastern Mediterranean Region, where a series of national seminars were held to bring all responsible persons from health and water agencies together to discuss the WHO Guidelines. Thus the way was paved for the drafting of national drinking water quality standards. This initiative is now followed up by national training courses for laboratory technicians who subsequently undertake the necessary analytical work. A training manual supports these courses and is available for similar courses in other countries (WHO 1988b).

In a joint effort with DANIDA, WHO regularly holds regional training courses which are attended by participants from neighbouring countries, as shown in Table 9.2 and Fig. 9.4. The particular target groups are district health inspectors and water authority personnel together with centrally responsible agency staff. All elements of drinking water quality control are covered from the planning stages to the application of preventive and remedial measures for the maintenance of safe water supplies. A training manual prepared jointly with the Technical University of Denmark supports these courses and provides for practical demonstration and interpretation of the WHO Guidelines in rural water supply situations (WHO/TUD 1988).

In another joint venture with UNEP and the Centre for International Projects, Moscow, WHO organizes multilingual workshops in the USSR which are attended by nationally responsible health and water agency staff from all continents. Information is exchanged and national experiences compared, and the results of recent pilot projects, technological developments, etc. are discussed at these workshops.

These global, as well as the regional seminars, training courses, workshops, etc. are important fora to present new techniques to verify the applicability of methods under varying hydrological and socio-economic conditions, and to stimulate innovative solutions to water quality problems. Thus not only the train-the-teachers concept is enhanced, but also the researchers and technologists receive their share of training in application and feasibility of new solutions to old problems of safeguarding water quality under rural area conditions. A review of the conceptual approaches to the training of national and district staff of health and water agencies will be undertaken jointly with previous participants in such training activities at a special WHO/DANIDA course seminar in Arusha, 1990.

Table 9.2 WHO/DANIDA training course venues

Mongu, Zambia	October	1986
Ouagadougou, Burkina Faso	November	1987
Khartoum, Sudan	November/December	1987
Yogyakarta, Indonesia	September/October	1988
Arusha, Tanzania	November	1989
Arusha, Tanzania	November	1990

FIG. 9.4 Course inauguration at Mongu, the first WHO/DANIDA training course on the control of drinking water quality in rural areas.

References

Barabas, S. (1986) Monitoring natural waters for drinking water quality. *World Health Statistics Quarterly*, **39** (1), 32–45.

Cvjetanovic, B. (1986) Health effects and impacts of water supply and sanitation. *World Health Statistics Quarterly*, **39** (1), 105–17.

Dahi, E. (ed.) (1989) *Environmental Engineering in Developing Countries – A Text Book*. Centre for Developing Countries, Technical University of Denmark, Copenhagen.

DelAgua Ltd. (1988) *Oxfam DelAgua Water Testing Kit – Users' Manual.* The Robens Institute, Environmental Health Unit, University of Surrey, Guildford.

Esrey, S. A., Feacham, R. and **Hughes, J.** (1985) Interventions for the control of diarrhoeal diseases among young children; improving water supplies and excreta disposal facilities. *Bulletin of the World Health Organization*, **63** (4), 757–72.

Feachem, R. *et al.* (1978) *Water, Health and Development*. Trimed Books, London.

Gibbs, K. R. (1983) Diarrhoea and the Decade in a developing country; the Bangladesh case. In *Proc. World Water Conference: The World Problem*, **21**, 189–95.

Government of Indonesia (1975) 01/BIRKHUMAS/1/1975 (01/legal Dept. Ministry of Health/1/1975 on drinking water quality). Government of Indonesia (1977) 173/Men. Kes Per/VIII/1977 (173/Ministry of Health regulation on water resources and waste water).

Green, D. M., Scott, S. S., Mowat, D. A. E., Sheerer, E. J. M. and **Thomson, J.M.** (1968) Water-borne outbreak of viral gastroenteritis and *Sonne* dysentery. *J. Hyg, Camb.* **66**, 383–92.

Helmer, R. (1989) Drinking water quality control: WHO cares about rural areas. *Waterlines*, **7** (3), January, 2–5.

Helmer, R. and **Ozolins, G.** (1987) Safeguarding water quality. In *World Water '86 – Water Technology for the Developing World.* Thomas Telford, London, pp. 25–30.

Hewison, K., Mac, K. F. and **Sivaboruorn, K.** (1988) *Evaluation of a Hydrogen Sulphide Screening Test.* Report No. 47, November. Thai Australian NE village water resource project. Australian Development Assistance Bureau, MPW Australia, 302 Little Lonsdale Street, Melbourne.

Institute of Child Health (1982) Report on action research on acceptability of safe water system and environmental sanitation by the rural communities of West Bengal. Calcutta (unpublished).

Krasovsky, G. (ed.) (1986) *Hygienic Criteria of Drinking Water Quality.* Centre for International Projects, GKNT, Moscow.

Lewis, W. J. and **Chilton, P. J.** (1984) Performance of sanitary completion measures of wells and boreholes used for rural water supplies in Malawi. *Proc. of the Harare Symposium*, July, 235–47.

Lewis, W. J. and **Chilton, P. J.** (1989) The impact of plastic materials on iron levels in village groundwater supplies in Malawi. *J. IWEM*, **3**, 82–8.

Lloyd, B. (1982) Water Quality Surveillance. *Waterlines*, **1**, **2**, 19–23.

Lloyd, B. (1988a) *Institutional Development of Rural Water Supply and Sanitation: Bengkulu and Lampung Provinces, Indonesia.* STC report with particular reference to water surveillance, 26 March–26 April, WHO project INO CWS 007.

Lloyd, B. (1988b) *Institutional Development of Water Supply and Sanitation with Particular Reference to Implementation of Water Surveillance in Indonesia.* WHO–STC assignment report, 22 Sept.–25 Oct., WHO project INO CWS 007. WHO–UNDP technical cooperation programme with the Government of Indonesia. UNDP project INS/85/031.

Lloyd, B. J., Pardon, M., Wedgewood, K. and **Bartram J.** (1986a) *Developing Regional Water Surveillance in Health Region XIII – Peru.* DelAgua, ODA Phase One Report, Peruvian Water Surveillance Programme.

Lloyd, B. J., Pardon, M. and **Bartram J.** (1986b) The development of a surveillance programme for Peru. *Proc. Int. Conference for Resources Mobilisation for Drinking Water and Sanitation in Developing Nations.* Am. Soc. Civ. Eng., Puerto Rico, pp. 640–52.

Lloyd, B. J., Pardon, M. and **Wheeler, D. C.** (1986c) *The Development Evaluation and Field Trials of Small Scale Treatment System for Rural Water Supplies.* Final report ODA research project R3760, DelAgua–Robens Institute, July.

Lloyd, B. and **Pardon, M.** (1988) The performance of slow sand filters in Peru. In *Slow Sand Filtration; 'Recent Developments in Water Treatment Technology'.* Ed. N. J. D. Graham, Ellis Horwood, pp. 393–411.

Lloyd, B., Pardon, M. and **Bartram, J.** (1989) Improving piped water supplies in Peru. *Waterlines*, **7** (3), 24–6.

Lloyd, B. and **Suyati, S.** (1989) A pilot rural water surveillance project in Indonesia. *Waterlines*, **7** (3), January, 10–13.

McJunkin, F. E. (1982) *Water and Human Health.* US Agency for International Development, Washington, DC.

Pardon, M. (1987) *Interim Evaluation of the WHO/UNEP Drinking Water Quality Surveillance Projects under Implementation in the Countries of Indonesia, Peru and Zambia.* May, Yerevan, USSR.

Pardon, M. (1989) The treatment of turbid water for small community supplies. Ph.D. thesis, University of Surrey.

Skilton, H. E. and **Wheeler, D.** (1988) Bacteriophage tracer experiments in groundwater. *J. Appl. Bact.*, **65**, 387–95.

Tschannerl, G. and **Bryan, K.** (eds) (1985) *Rural Water Supply Handpumps.* World Bank technical paper No. 48 and UNDP project management report No. 5, China.

Utkilen, H. and **Sutton, S.** (1989) Experience and results from a water quality project in Zambia. *Waterlines* **7** (3), 6–8.

Wardojo, R. S. (1987) *Evaluation of Water Monitoring Quality Control in Wonosari.* WHO report Jan–Feb, 1987. IND CWS 001/RB/1986.

Wimpenny, J. W. T., Cotton, N. and **Statham, M.** (1972) Microbes as tracers of water movement. *Water Research*, **6**, 731–9.

WHO (1976a) *Surveillance of Drinking Water Quality.* WHO monograph series No. 63. WHO, Geneva.

WHO (1976b) *Typical Designs for Engineering Components in Rural Water Supply.* WHO regional publications, South-East Asia Series No. 2, WHO/SEARO, New Delhi.

WHO (1981) *Drinking Water and Sanitation, 1981–1990: A Way to Health.* WHO, Geneva.

WHO (1984/85) Guidelines for Drinking Water Quality, Vol. 1, *Recommendations;* Vol. 2, *Health Criteria and Other Supporting Information;* Vol. 3, *Drinking Water Quality Control in Small Community Supplies.* WHO, Geneva.

WHO (1986) *Establishing and Equipping Water Laboratories in Developing Countries.* WHO document PEP/86.2, WHO, Geneva.

WHO (1987a) *Control of Environmental Health Hazards – a WHO Strategy for Technical Cooperation with Member States.* WHO document WHO/EHE/87.1, WHO, Geneva.

WHO (1987b) *GEMS/WATER Operational Guide.* WHO, Geneva.

WHO (1988a) *Towards the Target (an Overview of Progress in the First Five Years of the DWSSD).* WHO document WHO/CWS/88.2, WHO, Geneva.

WHO (1988b) *Training Course Manual for Water and Waste Water Laboratory Technicians.* WHO document WHO/PEP/88.11, WHO, Geneva.

WHO (1988c) *Introduction to National Seminars on Drinking Water Quality.* WHO document WHO/PEP/88.10, WHO, Geneva.

WHO (1988d) *Review of Progress of the International Drinking Water Supply and Sanitation Decade, 1981–1990: Eight Years of Implementation.* WHO document EB83/3, WHO, Geneva.

WHO (1989a) *UNEP/WHO Project on Control of Drinking Water Quality in Rural Areas.* WHO document WHO/PEP/89.5, WHO, Geneva.

WHO (1989b) *Revision of the WHO Guidelines for Drinking Water Quality.* Report of a consultation in Rome, Italy, 17–19 Oct. 1988. WHO document WHO/PEP/89.4, WHO, Geneva.

WHO (1989c) *UNEP/WHO Project on Control of Drinking Water Quality in Rural Areas.* WHO document WHO/PEP/8GJ, WHO, Geneva.

WHO/EURO (1986) *Standard Model Designs for Rural Water Supplies.* Prepared in cooperation with Hydroprojekt Consulting Engineers, Czechoslovakia. WHO Regional Office for Europe, Copenhagen.

WHO/TUD (1988) *WHO/DANIDA Course on Surveillance and Drinking Water Quality in Rural Areas.* Course manual for use in Yogyakarta, Indonesia, 1988. Centre for Developing Countries, Technical University of Denmark, Copenhagen.

Winter, J. A. (1985) Quality assurance support to the GEMS/WATER programme. *Water Quality Bulletin,* **10** (4), 181–5, 216.

WRC (1989) *Disinfection of Rural and Small Community Water Supplies – A Manual for Design and Operation.* Water Research Centre, Medmenham, England.

Index